Practical Geography
Presentation and Analysis

Practical Geography
Presentation and Analysis

Ken Briggs

HODDER AND STOUGHTON
LONDON   SYDNEY   AUCKLAND   TORONTO

Copyright © 1989 Ken Briggs

British Library Cataloguing in Publication Data
Briggs, K. (Kenneth), 1922–
    Practical geography: presentation and analysis.
    1. Geographical features
    I. Title
    910

ISBN 0 340 41567 3

All rights reserved. No part of this publication may be reproduced or transmitted in any form or by any means, electronically or mechanically, including photocopying, recording or any information storage or retrieval system, without either the prior permission in writing from the publisher or a licence permitting restricted copying. In the United Kingdom such licences are issued by the Copyright Licensing Agency, 33–34 Alfred Place, London WC1E 7DP.

Typeset in Linotron Times by Photographics, Honiton
Printed in Great Britain for Hodder and Stoughton Educational, a division of Hodder and Stoughton Ltd, Mill Road, Dunton Green, Sevenoaks, Kent, by Thomson Litho Ltd, East Kilbride

Contents

1	**Graphs**	**1**
	Histograms, frequency polygons and ogives	2
	Bar charts and circular graphs	6
	Line graphs	11
	Scatter diagrams	26
	Best-fit line	30
2	**Maps**	**36**
	Dot maps	36
	Choropleth maps	37
	Proportional symbols	41
	Isoline maps	44
	Flow diagrams	50
3	**Matrices**	**57**
4	**Techniques Used in the Analysis of Data**	**61**
	Averages	61
	Measures of dispersion	65
	Skewness in a distribution	68
	The normal distribution	68
5	**Techniques Used in the Analysis of Spatial Point Patterns**	**71**
	Averages	71
	Measures of spatial dispersion	73
6	**Techniques Used in the Analysis of Networks**	**81**
	Drainage networks	81
	Transport networks	84
7	**Specialisation—the Location Quotient**	**92**
8	**The Spearman Rank Correlation Coefficient**	**95**
9	**Weather Maps**	**99**
10	**Topographical Maps**	**105**
	Aspects of physical geography	105
	Aspects of human geography	109
11	**Photographs**	**112**
	The interpretation of aerial photographs	112
	The interpretation of ground-level photographs	113
	Index	**121**

Preface

The aim of this book is to present a variety of practical techniques, which are commonly specified in Advanced Level syllabuses, in a way that can be readily understood. This objective is achieved by presenting explanatory material clearly and by linking it with a large number of varied diagrams. The book enables students to teach themselves without having to seek assistance from their teacher. The exercises at the end of each section of the book are designed to consolidate students' understanding of the foregoing explanation. The practical techniques that are explained in the book are applicable to examination papers in practical geography and also to theoretical examination papers, as well as the processing of information derived from fieldwork.

The book is concerned with the construction and interpretation of graphs, maps and matrices and the analysis of quantitative and spatial data. Together with *Physical Geography* and *Human Geography* by the same author, it constitutes a sound, balanced course in Advanced Level Geography.

K. Briggs

Acknowledgements

The following sources were used for some of the diagrams and data in this book.

Monthly Digest of Statistics CSO, February 1986 (1.6 (a)),
The Geographical Digest (George Philip), 1984 (1.8),
'Shifts in US coal production: Trends and implications'. Table 1 *This Changing World (Geography)* November 1981 (1.9),
Pearce, E. A. and Smith, C. G., *World Weather Guide* Hutchinson 1984 (1.12),
'West German dilemma : Little Turks or Young Germans' : *This Changing World (Geography)* April 1984 (1.13),
Harris, C. D., Urban and Industrial Transformation of Japan *Geographical Review* January 1982 (1.27),
The Geographical Digest (George Philip), 1984 (2.1, 2.2, 2.3),
Clem, R. S., Regional Patterns of Population Change in the Soviet Union 1957–79 *The Geographical Review* April 1980 American Geographical Society (p. 94),
Bimal KantiPaul, Malaria in Bangladesh *The Geographical Review* American Geographical Society January 1984 (p. 98),
Bolton Evening News (11.2).

1 Graphs

A graph is a diagrammatic or pictorial representation of the characteristics of a set of variables. Variables are characteristics that can vary numerically, such as temperature, rainfall, exports, population totals, examination marks and cricket scores. Graphs are of great value because they help us to understand large

(a) *Raw data*

34
26
31
18
13
14
30
16
28
21
10
12
20
27
28
16
14
13
23
10
32
31
10
8
19
14
16
25
16
34

(b) *Simple array*

8
10
10
10
12
13
13
14
14
14
16
16
16
16
18
19
20
21
23
25
26
27
28
28
30
31
31
32
34
34

(c) *Array with frequencies*

Altitude	Frequency
8	1
10	3
12	1
13	2
14	3
16	4
18	1
19	1
20	1
21	1
23	1
25	1
26	1
27	1
28	2
30	1
31	2
32	1
34	2
	30

(d)

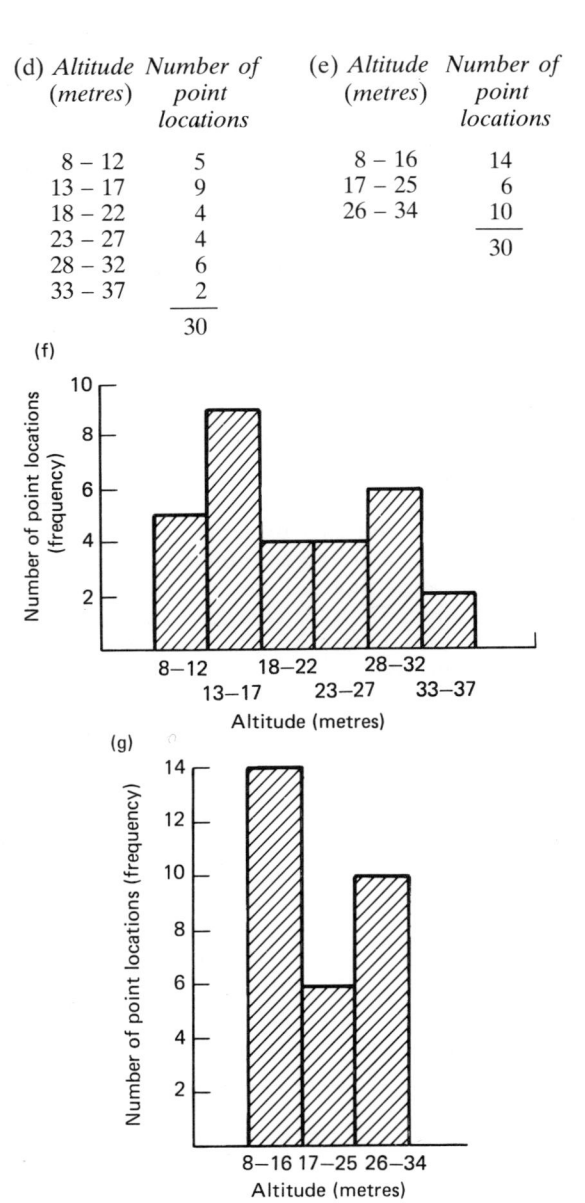

Altitude (metres)	Number of point locations
8 – 12	5
13 – 17	9
18 – 22	4
23 – 27	4
28 – 32	6
33 – 37	2
	30

(e)

Altitude (metres)	Number of point locations
8 – 16	14
17 – 25	6
26 – 34	10
	30

(f)

(g)

Fig. 1.1 Sample of altitudes (metres) above sea level for 30 point locations

amounts of complex data. Such data are organised and summarised by a graph. Also, a graph can make clear the similarities and differences between two or more sets of complex data.

In constructing a graph we first of all collect items of numerical data and arrange them in a list in the order in which they occur (Fig. 1.1(a)). This list can be referred to as 'raw data'. It has not been processed in any way. It is clearly of very little value because it gives no indication of the numerical pattern formed by the set of numbers. One can deduce only the broadest of generalisations about its characteristics. A simple way to improve our understanding of the data is to arrange the numbers into either an ascending or a descending order of magnitude. Such a list is referred to as an 'array' (Fig. 1.1(b)). This provides us with some easily accessible information. For example, we can easily identify the largest and smallest values and we can therefore calculate the difference between them (i.e. the 'range' of the data).

By looking down the list of numbers we can also learn something about the variations that exist between the two extreme values. These are set out as a frequency distribution in Fig. 1.1(c). However, it is not easy to gain an overall understanding of the characteristics of the sample because there are too many individual items of information. We therefore summarise the data into six equal ranges of altitude and allocate each frequency to its appropriate range (Fig. 1.1(d)). In so doing we have sacrificed some of the detailed information in order to improve our understanding of the overall picture. We can now understand the data and we should be able to make comparisons with other areas quite easily. It is however undesirable to carry this process too far. Figure 1.1(e) shows the undesirable effect of grouping the data into only three ranges of altitude. The 8m – 16 m group includes nearly half the observations. Figure 1.1(f) and Figure 1.1(g) are simple examples of graphs.

1.1 Histograms, frequency polygons and ogives

HISTOGRAMS

One type of graph that illustrates a frequency distribution is a histogram. This consists of a set of vertical bars or rectangles rising from the graph's X axis, as in Fig. 1.1(f) and (g). The Y axis is labelled as frequency. The histogram is constructed in such a way that the *areas* of these rectangles are proportional to the class frequencies, as was the case in Figure 1.1(f) and (g). In a histogram however the *width* of the rectangle is permitted to vary. Hence, if one rectangle is twice the width of each of the others, it will have to have half the height in order to maintain the correct relationship between the rectangle's area and the class frequency. In a histogram the X axis indicates the value of the *independent variable*, as is the general rule. The Y axis indicates the value of the *dependent variable*, i.e. *frequency*.

The width of each rectangle represents the width of the class between its upper and lower class limits. This is referred to as the *class interval*. Gaps should never be left between the rectangles. Class intervals, if possible, should be equal along the whole of the X axis so as to avoid the 'area' problem mentioned above. A histogram should have at least five class intervals (rectangles), but it should not have so many that zero frequencies begin to appear.

The boundaries of the various classes along the X axis are referred to as *class boundaries*. A problem can arise concerning their precise location. It is essential that no frequency value is either omitted or counted twice. The solution depends on the type of data that is being processed. One type of data is termed '*discrete*'. This consists of the enumeration (counting) of units of a variable. This kind of data must increase or decrease by at least one whole unit. An example is the number of children in a family. It is extremely well suited to being used in histograms because there is no difficulty in determining class intervals and boundaries. Data

GRAPHS

can be arranged in class intervals of, for example, 1 – 10, 11 – 20, 21 – 30 etc. No values can exist between 10 and 11, 20 and 21 etc. *Continuous* data on the other hand, is obtained by measurement and it can be accurate to several places of decimals. Nevertheless it can still be used in histograms. A class may be defined as extending from 10 to 19.9999, this being, in effect 20. But the class boundaries must not overlap, so the class is said to extend from 10 to 'less than 20'. In cases in which the data has been recorded to the nearest centimetre or metre, a class described as '20 – 24m' could be regarded as extending from 19.5m to 24.5m.

It is not usual to label a histogram with the value of the class boundaries, as shown in Figure 1.1(*f*) and (*g*). Instead the normal practice is to label the *class mark* beneath the X axis at the centre of the base of each rectangle. The class mark is the mid-point of the class interval. It is calculated by adding the value of the upper class boundary to that of the lower class boundary and then dividing by 2. For *discrete data*, class intervals of 20 – 24, 25 – 29, 30 – 34 respectively give class marks of 22, 27 and 32. For continuous data a class interval of 10 to less than 20 gives a class mark of 10 + 19.9999 divided by 2. This is virtually the same as 10 + 20 divided by 2, which equals 15.

If the values represented in the histogram are to be given further processing, it is not possible to recall the original numbers which were combined into classes. Hence, all values occurring within a given class interval are assumed to be equal to the class mark. This, of course, is likely to produce a slight error.

The construction of a histogram

Figure 1.2 illustrates the way in which a simple histogram is constructed. Two different histograms with different class intervals have been constructed, using the same data. In all cases the class mark is located centrally within the class interval and in each graph the class intervals are the same size. The values for the birth rates have been rounded off so that there are no fractions, but, for example, the class with a class mark of 14 in Histogram A can be regarded as extending from 12.5 (midway between 11 and 14) to 15.5 (midway between 14 and 17).

It is interesting to test the accuracy of Histogram A against the original values of the birth rate given in Figure 1.2(*a*). This can be done by multiplying the histogram's class mark by its frequency and comparing this value with the sum of the corresponding birth rates in the array. This comparison is tabulated below.

Birth Rate	Sum of birth rates in array	Histogram value
10 – 12	78	77
13 – 15	182	182
16 – 18	49	51
19 – 21	99	100
22 – 24	24	23
25 – 27	25	26

It appears from this table that Histogram A reflects the original data reasonably accurately despite the rounding that has taken place. Histogram B, however, is slightly less accurate, especially in the large class whose class mark is 12.

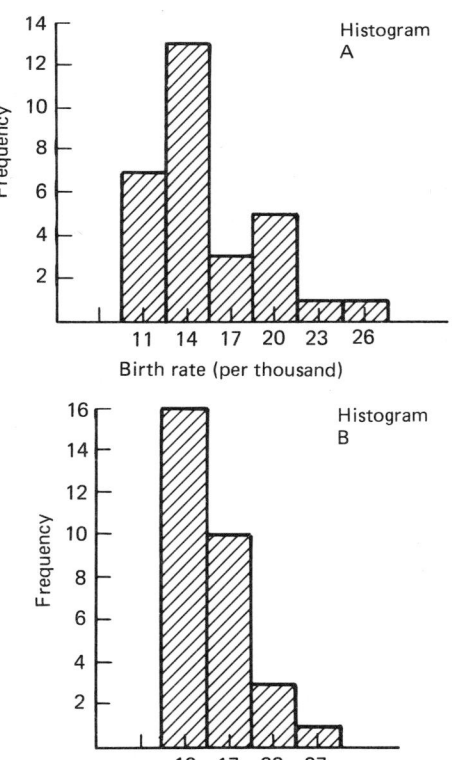

Fig. 1.2 The construction of a histogram

In the table below is recorded a sample of national birth rates (per thousand persons) for countries in the northern hemisphere. The table is set out as an array.

10 10 11 11 12 12 12 13 13 13 13 14 14 14 14 14
15 15 15 15 16 16 17 19 19 19 21 21 24 25

The highest birth rate is 25 and the lowest is 10, so the *range* of the data is 15. If we arrange the birth rates into 3 groups the size of the class intervals will be about 5. If we arrange the birth rates into 5 groups the size of the class intervals will be about 3.

Histogram A Class interval 3		Histogram B Class interval 5	
Birth rate	*Frequency*	*Birth rate*	*Frequency*
10 – 12	7	10 – 14	16
13 – 15	13	15 – 19	10
16 – 18	3	20 – 24	3
19 – 21	5	25 – 29	1
22 – 24	1		30
25 – 27	1		
	30		

Fig. 1.2(a) Data for the construction of a histogram

The shape of a histogram

A histogram's shape is described by referring to three different characteristics formed by its rectangles.

One of these characteristics is *skewness*. This refers to the balance or symmetry of the histogram's outline. If, instead of being symmetrical, the histogram has a 'peak' (the modal class) to the left of its centre and a 'tail' extending towards the right, it is said to have positive skewness. In this case the 'tail' extends towards the higher values of the variable. A histogram with a 'tail' towards the left and a 'peak' towards the right is negatively skew. Both of the histograms in Figure 1.2 are positively skew.

Another characteristic concerns the differences between the frequencies of the various classes. The class with the highest frequency (whose rectangle is longest) is called the modal class. If a histogram has a single modal class and frequencies decline steadily to its left and right, it is said to be unimodal. Histogram B in Figure 1.2 is an example. In some cases there may be two classes that rise above the rest, as in Histogram A. Such a distribution is called bimodal. If there are several 'peaks' the distribution is multimodal.

Kurtosis is another term used to describe a histogram's shape. This refers to the differences in the heights of the rectangles. A histogram with rectangles fairly equal in height is referred to as platykurtic. One with a high 'peak' is leptokurtic. A generally average shape, between these two extremes is mesokurtic. These terms also apply to other types of frequency diagrams in addition to the histogram.

FREQUENCY POLYGONS

The construction of a frequency polygon
A frequency polygon is very similar to a histogram. Its Y axis shows frequency and its X axis the class mark. To construct the polygon from a histogram, straight lines are drawn linking the mid-points of the tops of the histogram's rectangles (Fig. 1.3). This means that the frequency polygon encloses exactly the same area as the histogram. In effect, the histogram has been changed into a simple line graph. A frequency polygon can of course be constructed without first drawing a histogram. All that is necessary is to plot frequency against each class mark and then join up the dots. But at each end of the graph the frequency polygon must be extended beyond the first and last classes (Fig. 1.3).

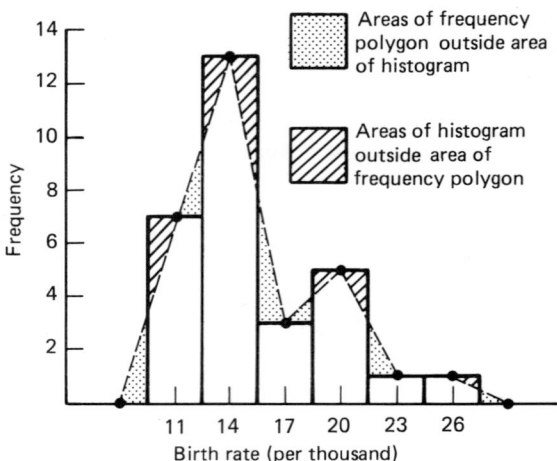

Fig. 1.3 The relationship between a frequency polygon and a histogram

GRAPHS

In Figure 1.3 it can be seen that, working from left to right each dotted triangle outside the histogram is next to a shaded triangle of exactly the same size and shape inside the histogram. Hence the total area of the dotted triangles is equal to that of the shaded ones. A frequency polygon can be described in the same terms (skewness, kurtosis, etc.) as a histogram.

Sometimes a frequency polygon is smoothed to form a frequency curve. The use of such a curve can lead to considerable inaccuracy unless there are a large number of classes with very small class intervals.

OGIVES

The construction of an ogive

An ogive shows the cumulative frequency of a distribution. The cumulative frequency is the total frequency of all the values that are less than the upper class boundary of any class interval. In the cumulative frequency table for Histogram A in Figure 1.4 there are 7 values less than 12.5 and 13 values of at least 12.5 but less than 15.5. So the cumulative frequency of values less than 15.5 is 7 + 13 = 20. It should be noted that, in calculating the cumulative frequency, the class marks of the histogram are not used. The numbers on the X axis of the histogram are the values of the class boundaries. The shape of the ogive is related to the shape of its corresponding histogram in the following

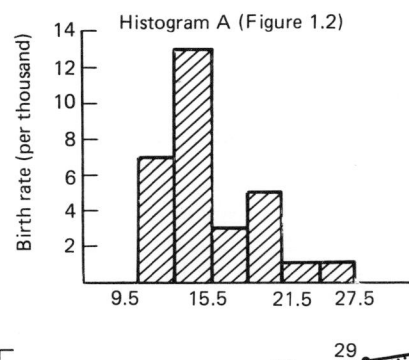

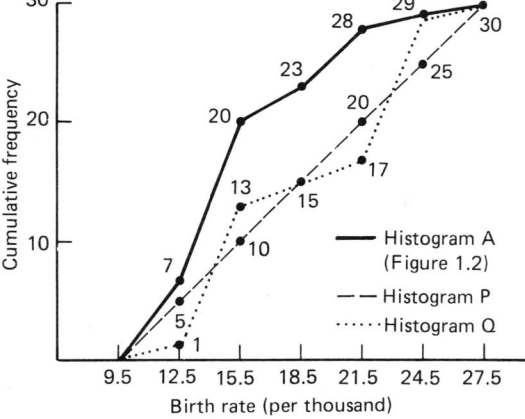

The numbers indicated on the X axis are the class boundaries and not the class marks.

Cumulative frequency table (from the above graph)

Class range	Class frequency	Cumulative frequency
Less than 9.5	0	0
9.5 – under 12.5	7	7
12.5 – under 15.5	13	20
15.5 – under 18.5	3	23
18.5 – under 21.5	5	28
21.5 – under 24.5	1	29
24.5 – under 27.5	1	30

Fig. 1.4 The construction of an ogive from a histogram

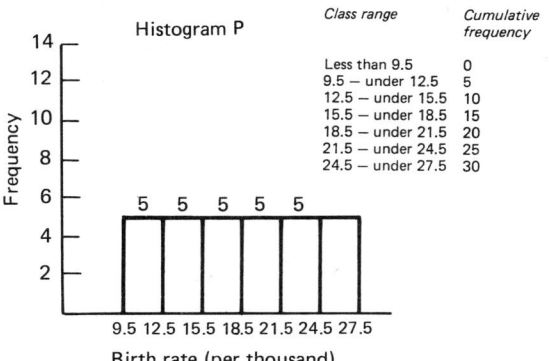

Class range	Cumulative frequency
Less than 9.5	0
9.5 – under 12.5	5
12.5 – under 15.5	10
15.5 – under 18.5	15
18.5 – under 21.5	20
21.5 – under 24.5	25
24.5 – under 27.5	30

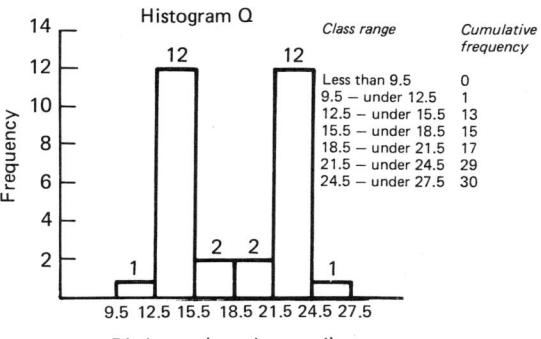

Class range	Cumulative frequency
Less than 9.5	0
9.5 – under 12.5	1
12.5 – under 15.5	13
15.5 – under 18.5	15
18.5 – under 21.5	17
21.5 – under 24.5	29
24.5 – under 27.5	30

ways. A very high class frequency in the histogram causes the ogive to rise steeply, as in the ogive for Histogram A (Fig. 1.4) between the birth rate values of 12.5 and 15.5. On the other hand, near the upper end of Histogram A's ogive the line is almost horizontal, showing that class frequencies here are very small. Histogram P (Fig. 1.4) illustrates a 'rectangular' distribution in which all classes of the birth rate have the same frequency. This produces an ogive in the shape of a straight line. Histogram Q, in which the frequency of the classes varies considerably, produces an ogive in the form of a zigzag line.

Care has to be exercised if you wish to compare two different ogives. It is not possible to make comparisons between ogives by using 'raw' frequencies unless the totals of their frequencies are equal. If the total frequencies are not equal a comparison can be made if, for both distributions, the frequencies are expressed as percentages of the total frequency.

1.2 Bar charts and circular graphs

BAR CHARTS

The construction of a bar chart
A bar chart or bar graph resembles a histogram in that it often consists of a series of vertical rectangles rising from the graph's horizontal axis. However, it differs from a histogram in certain important ways. In a bar chart the width of the bars is not important. All the bars are the same width and it is usual to leave a space between neighbouring bars except in the case of rainfall graphs. The *length* of the bar, and not its *area*, is proportional to the quantity represented.

In geography, the horizontal axis of the bar chart is usually divided into zones representing either 'areas', such as countries or towns, or 'time intervals' such as the months of the year or a succession of different years. The vertical axis has an arithmetic scale of measurement which can represent 'areas', population totals, production levels, population density and many other measured characteristics relating to the areas or the time periods indicated along the horizontal axis.

All bar charts should begin at the zero line on the vertical axis. If the bars are approximately the same length it may be desirable to emphasise the slight differences that exist by 'breaking' the vertical scale by means of a zigzag line near the bottom of the graph (Fig. 1.5(a)). It is *always* necessary to indicate the position of 'zero' on the vertical axis.

The characteristic represented by the rectangles may have several constituent parts for which quantitative data is available. The graph can be arranged so as to show a comparison between two or three sets of values. In this case it is possible to arrange the rectangles in groups for each area or date indicated on the horizontal axis. Figure 1.5(b), a multiple bar chart, is an example of this. There should never be more than three elements in each group of a multiple bar chart.

Alternatively, instead of grouping the rectangles side by side, the total length of the single vertical rectangle can be subdivided into three groups as in Figure 1.5(c) a component or compound bar chart. Figure 5(b) conveys information concerning the sub-groups very efficiently but one cannot visualise the changes in the production totals. Conversely, Figure 1.5(c) successfully indicates the variations in total employment but is weaker in showing the comparative values of the sub-groups.

In a compound bar chart it is usual to arrange the pattern so that the most stable element occurs at the bottom of each rectangle and the most variable one at the top. This is done to prevent a variable element from transmitting its variations upwards through the other elements above it. A stable element will cause little disturbance to the elements above it.

Figure 1.5(d) shows the varying *percentages* of three classes of employment over a five-year period. This graph has to be used with special care. Since it only shows percentages it gives no information about absolute totals. The total length of each bar represents the total employment for each year, but we are given no information about how the employment totals varied between the years. Hence, 50% of employment in 1981 does not necessarily equal 50% of employment in 1985. In some cases it

GRAPHS

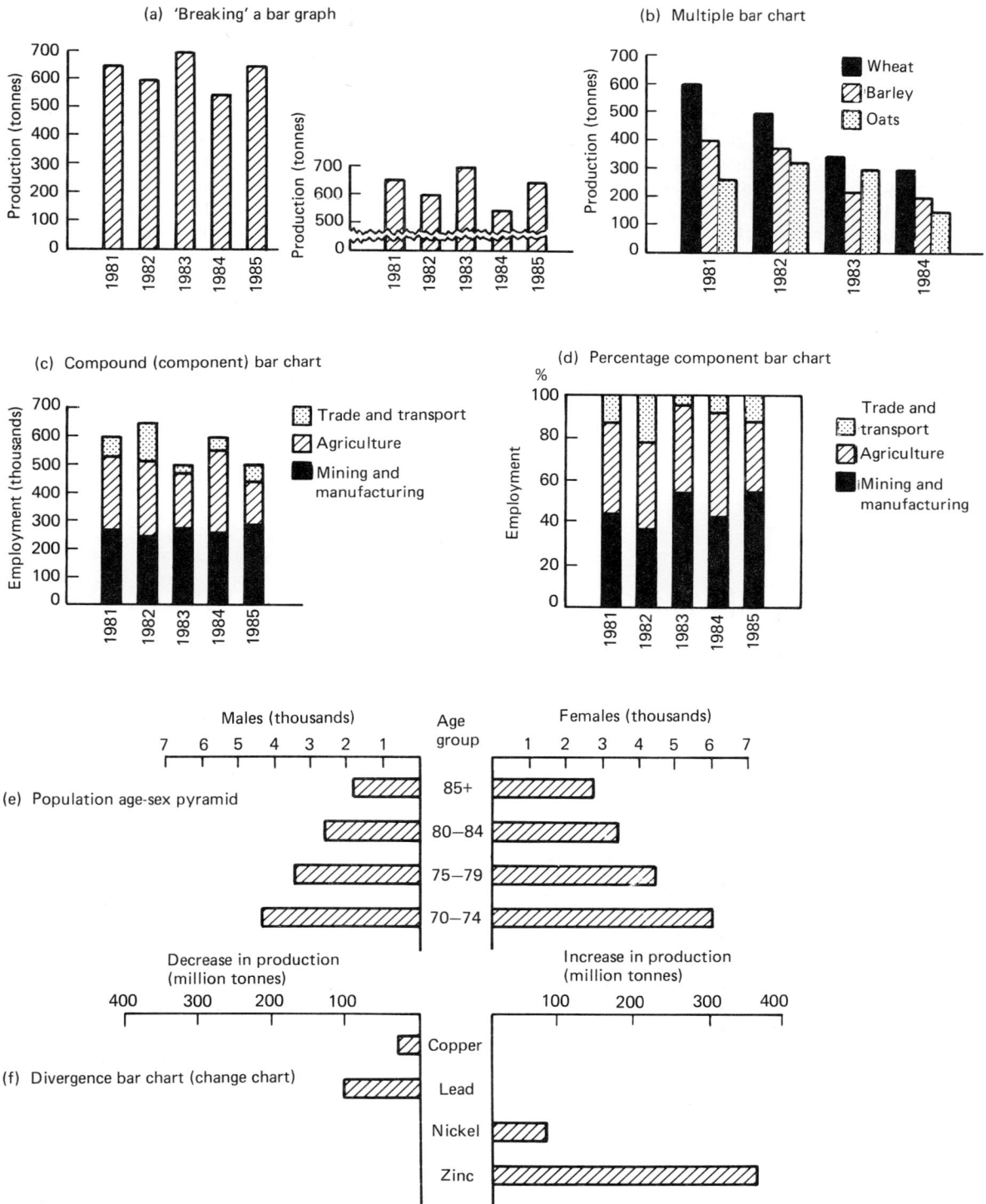

Fig. 1.5 Bar charts

is customary to align the bars of a bar graph horizontally. This is so in the case of a population age–sex pyramid. It is clear from Figure 1.5(e) that problems would arise if the diagram were rotated through 90° and the horizontal axis had to accommodate nearly 20 separate age ranges.

If a graph represents both increases and decreases of an element it is usually best to align the bars horizontally so that the rectangles representing increases can extend to the right of the zero axis and the decreases can extend to the left. This is consistent with the convention that, along the X axis of a graph, values of a variable increase towards the right. Figure 1.5(f) is an example of such a 'divergence bar chart' or 'change chart'.

Figure 1.6 illustrates in detail the differences between a compound bar chart and a multiple bar chart. Each of these emphasises a different type of information. In Figure 1.6(a) the compound bar chart makes it easy to understand the variations in total unemployment that have occurred (the total length of each bar). It is also easy to understand the variations in male unemployment. On the other hand it requires a little processing to identify the variations in female unemployment. These cannot easily be recognised from a glance at the graph. Figure 1.6(b) on the other hand gives a perfectly clear description of the variations in both male and female unemployment, but it is more difficult to identify the trends in total unemployment.

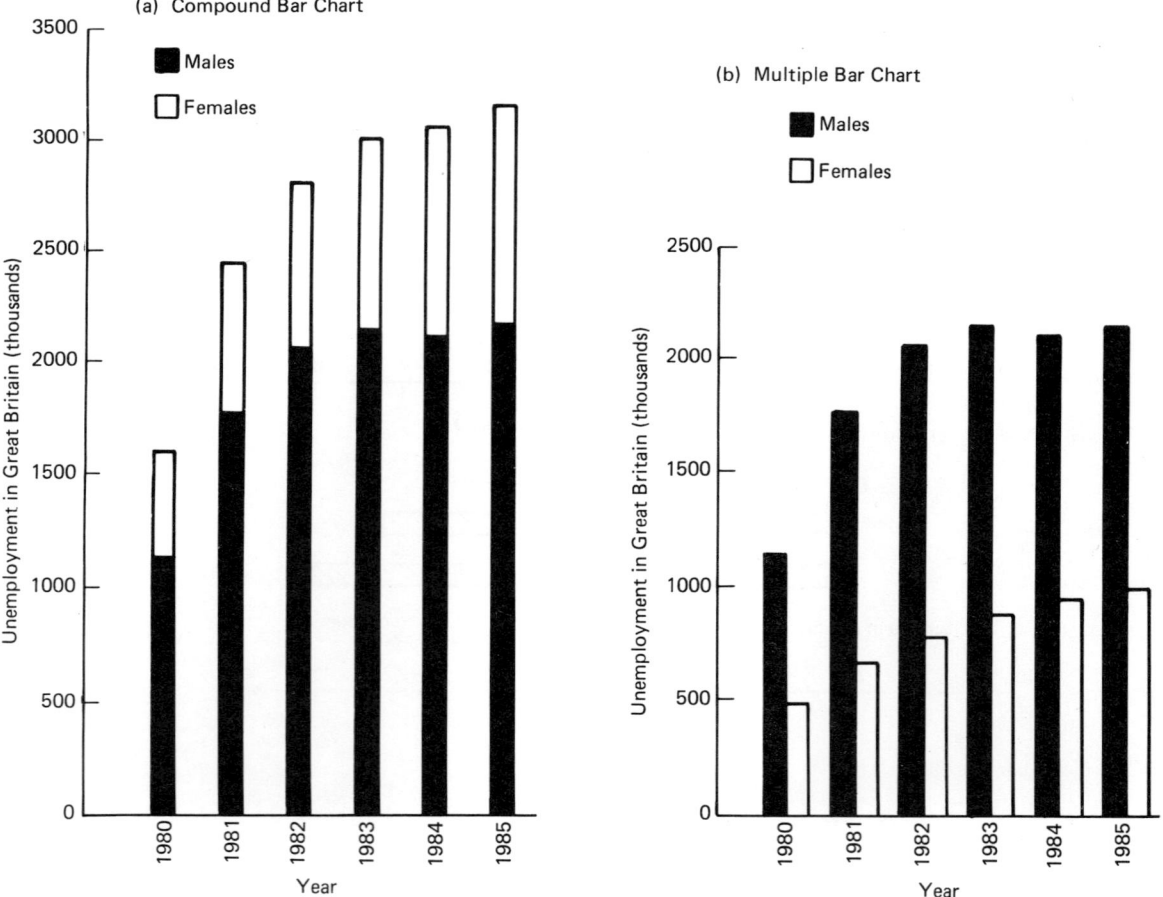

Fig. 1.6 Bar charts showing unemployment in Great Britain (1980–85)

GRAPHS

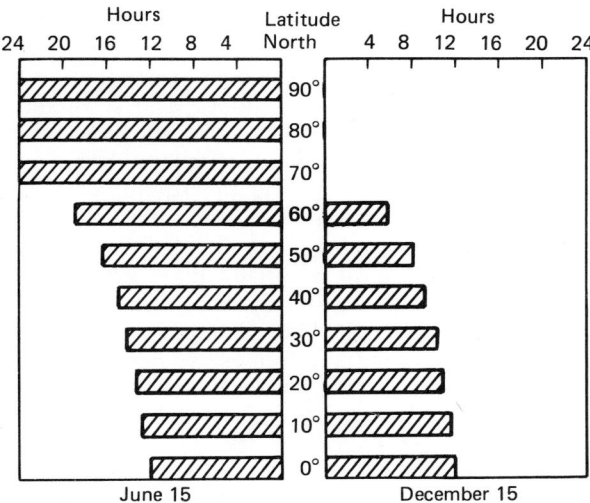

Fig. 1.7 Bar chart showing maximum possible hours of sunlight at various latitudes in the northern hemisphere

Figure 1.7 is very similar to the population age–sex pyramid but, instead of numbers of males and females in various age groups it shows the number of hours of sunlight at various latitudes in the northern hemisphere. Bar graphs of a particular type can be adapted to illustrate very different types of data.

CIRCULAR GRAPHS

The construction of circular graphs

There are two main types of circular graph. One of these is the *clock graph* or *polar graph* in which numerical values are indicated on a scale extending from the centre of the circle towards its perimeter. The other is the *pie graph* or *divided circle* in which there are two scales, one extending from the centre outwards and the other extending around the perimeter.

Examples of pie graphs are given in Figures 1.8 and 1.9. In this type of graph the area of the circle is made proportional to the total quantity stated in the data. The circle is also divided by radial lines into segments which are made proportional in size to the component parts of the data.

In constructing a pie graph a circle is drawn which is proportional in area to the total quantity

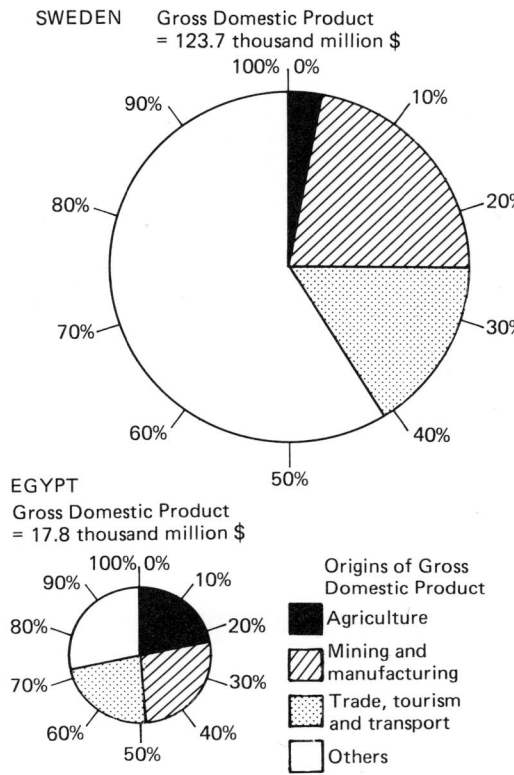

Fig. 1.8 The construction of a pie graph from 'percentage data'

that is to be represented. This is done by finding the square root of the total quantity and making this the radius of the circle in either millimetres or centimetres or some other unit according to the space available.

A separate circle is drawn for each set of values. The circle is then divided up into sectors which are proportional in size to the numbers represented. Ideally there should be at least 4 sectors but not more than 10. This process can be somewhat difficult and particular care needs to be taken with the smaller, narrow sectors. If possible it is best to use polar co-ordinate graph paper on which a hundred angles of 3.6° each are marked off at the centre of a circle. A pie graph can be constructed directly from either 'percentage data' (Fig. 1.8) or 'raw data' (Fig. 1.9).

Figure 1.8 is a pie graph which provides a broad comparison between the economies of Sweden and Egypt. First of all the two circles were drawn so as to be proportional in area to

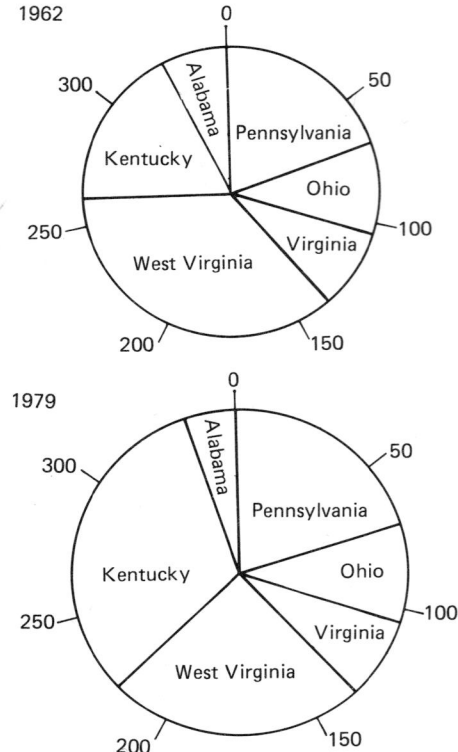

Fig. 1.9 The construction of a pie graph from 'raw data': bituminous coal production (million tonnes) in Appalachia (1962 and 1979)

the Gross Domestic Product of each country. The circle for Sweden is about seven times as large as that for Egypt. Next the circles were divided in proportion to the four economic sectors from which the Gross Domestic Product originated. The original data provided information as percentages. Since 100% on the graph is represented by a complete circle (360°) each 1% of the Gross Domestic Product is represented by an angle of 3.6° (3.6° × 100 = 360°). Hence, for example, Sweden's 'Agriculture' sector provided 3% of the Gross Domestic Product, so this is represented by an angle of 3 × 3.6° (i.e. 10.8°). All the other segment angles were calculated similarly. The sectors were then inserted into the circles and a scale of percentages was marked on the perimeter of each circle.

The two pie graphs show clearly the contrasts between the Swedish and Egyptian economies, but it would be necessary to process the data further in order to obtain *actual* values.

Figure 1.9, on the other hand, deals with actual values of bituminous coal production in Appalachia (USA). Here a much greater amount of calculation has to be done. The left-hand column of the data gives the actual total weight of bituminous coal produced in Appalachia in 1962 and 1979. By finding the square root of each of these totals (329.5 and 448.8) we can draw the two circles at their correct relative size. We next have to divide these circles. In 1962, Pennsylvania produced 65.3 million tonnes of bituminous coal. The whole of Appalachia produced 329.5 million tonnes. So Pennsylvania produced 65.3/329.5 (65.3 divided by 329.5), about one-fifth of Appalachia's total production. Pennsylvania's sector on the pie graph must therefore occupy about one-fifth of 360°, that is, about 70°. The actual figure is 71.3° (Figure 1.9). The remainder of the calculations are carried out in the same way. On the perimeter of each pie graph a scale (in million tonnes) has been marked, starting at 12 o'clock and increasing in a clockwise direction. Comparisons can now be made between the two pie graphs. It is clear that total production had increased considerably by 1979 (compare the sizes of the two pie graphs). The changes in the production levels of the individual states can be estimated from the two graphs. Kentucky, Alabama and Pennsylvania have clearly increased their production considerably. It is more difficult to decide in the cases of Ohio, Virginia and West Virginia.

Although the values cannot easily be read directly from the graph a pie graph gives a good general impression of differences and similarities. The pie graph is more difficult to construct than the bar graph but it does give a better general impression of the relationship between the various subdivisions and the overall total.

Pie graphs can also be useful in compiling statistical maps. Small pie graphs can be inserted within the outlines of countries or other administrative areas to describe some of their characteristics. Being circular the pie graph has a more compact shape than a bar graph and so is able to be fitted into comparatively small spaces.

Polar charts differ from pie graphs in that their scale of values runs radially outwards from the centre of the graph towards the perimeter instead of running along the perimeter. Thus,

GRAPHS

different values of the variable are plotted in the form of radii from the origin at the centre of the graph. Usually the lowest value shown in the scale is placed a short distance outwards from the origin (Figs. 1.10 and 1.11). The perimeter of the chart is divided regularly into the categories to which the scale values refer. These categories are commonly either the months of the year or the hours of the day. Twelve subdivisions of the perimeter produce sectors of 30° each and 24 divisions produce sectors of 15° each. Polar charts are particularly suited for illustrating a characteristic that tends to recur in successive time periods, such as mean monthly temperatures and rainfall.

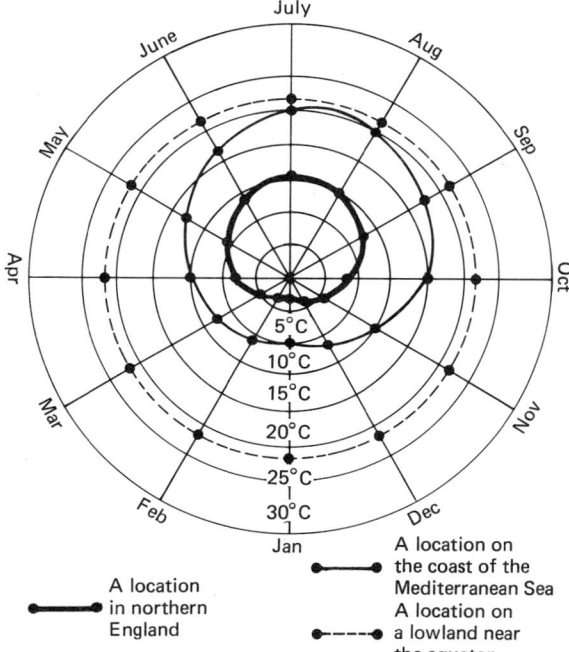

Fig. 1.10 A circular graph illustrating temperature regimes

Figure 1.10 illustrates the succession of mean monthly temperatures at three different locations. The use of the circular format dispels any tendency to regard January as the beginning of the year and December as the end. The desirable impression is given of a reasonably steady variation of temperature through the year. The temperature characteristics of these three locations have also been illustrated quite clearly, with no tendency for the graph to become crowded. Mean monthly rainfall totals are not so easy to illustrate by means of a graph if one observes the convention that amounts of rain are represented by a series of bars instead of lines. The style of Figure 1.11 does not permit more than one location to be represented on a single graph without risking congestion.

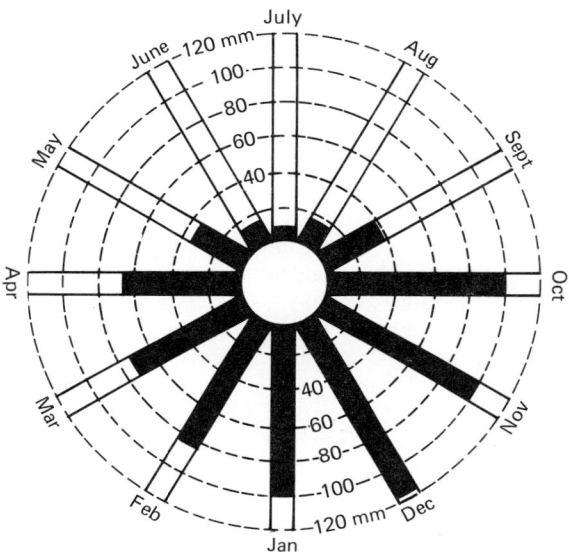

Fig. 1.11 A circular graph illustrating the rainfall regime at a location on the coast of the Mediterranean Sea

In a wind rose, instead of the months of the year, the perimeter of the circle is divided into sectors representing either 8 (N, NE, E, SE, etc.) or 16 (N, NNE, NE, ENE, etc.) points of the compass. Radiating lines or bars indicate the comparative frequency or strength of winds *from* these various directions.

1.3 Line graphs

A simple line graph shows how one variable changes in relation to changes in another variable, which is frequently 'time'. Figure 1.10, a circular graph, is in fact a type of line graph because it shows how temperatures change in relation to the months of the year. In Figure 1.10, however, the line has been drawn through points located in relation to a scale marked along the circumference of a circle and another scale along lines radiating from the centre of the circle. In this section however we are concerned with graphs which have a horizontal X axis and a vertical Y axis. The X axis shows values of the independent variable that change steadily. The Y axis shows values of the dependent variable that often change more erratically. The Y axis is labelled with quantitative or percentage values of variables such as population totals, the value of imports and exports, levels of production for various commodities and temperature. For example, temperatures (dependent variable) can be plotted against months of the year, altitude or latitude (independent variables).

The vertical scale should always start at zero but, if there is a considerable gap along the axis before the lowest value is reached the vertical scale can be interrupted, as in Figure 1.5(*a*). In planning the line graph it is important to note the maximum value of the dependent variable (Y axis) so that a vertical scale that neatly fits the data can be drawn. On the vertical scale the values should be kept as concise as possible. For example, instead of labelling the intervals '20 000 000' and '30 000 000' it is better to label 'millions' at the top of the scale and quote the interval values as '20' and '30'. Items of the data should be plotted on the graph as a series of dots and finally the dots should normally be linked with a series of straight lines.

The construction of a simple line graph

Line graphs are particularly easy to construct. All that is necessary is to plot a series of dots on the graph paper in the correct locations and then draw straight lines between adjacent dots. Figure 1.12 is an example of a *multiple line graph* (or polygraph). Four different values of temperature are indicated for each month. This gives a much more informative summary of the temperature regime than a single line showing only the mean monthly temperatures. Data such as this can also be displayed on a circular graph like Figure 1.10. Figure 1.13 is a *compound line graph* which, in addition to showing the total inflows of foreign labour to West Germany, also gives details of the labour's country of origin. This kind of graph is rather more difficult to construct than a multiple line graph. Not only

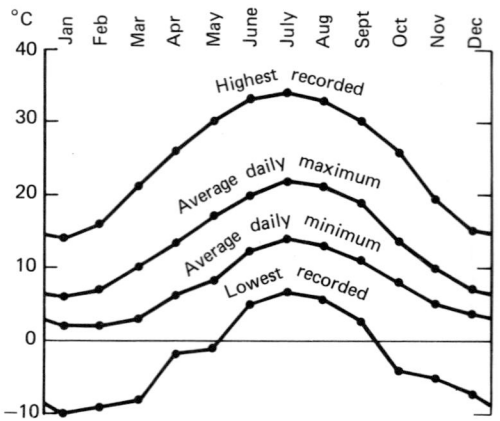

Fig. 1.12 Multiple line graph (polygraph) showing temperatures in London

have the vertical measurements to be correct for each of the subdivisions (Australia, Turkey, etc.) but also the sum of all these must be equal to the total inflow in any year. It is probably easiest to plot first the category that is placed lowest on the graph (e.g. 'Other'). The next dot should be placed at the value which represents the combined totals for 'Other' and Yugoslavia, then the combined totals for 'Other', Yugoslavia and Turkey, and finally the grand total.

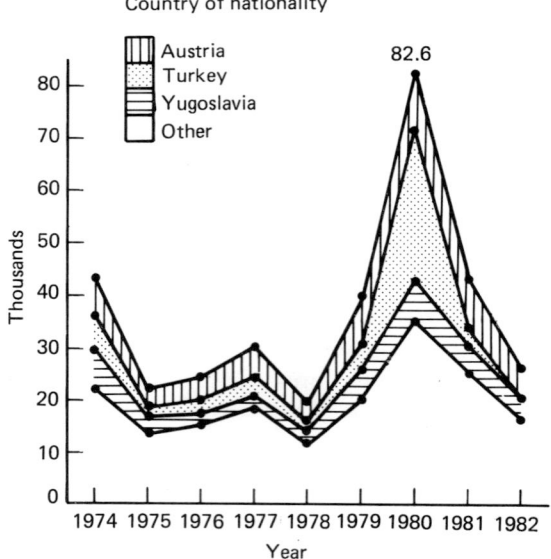

Fig. 1.13 Compound line graph showing inflows of foreign labour to West Germany (1974–82)

GRAPHS

The smoothing of a line graph

Sometimes a simple line graph can show quite violent fluctuations in its values. An example is the very high peak in 1980 for the inflow of Turkish labour to West Germany. In other cases there may be variations in the production of a commodity from year to year or there may be unusual variations in temperature between the months of a year. Such variations may make it difficult to understand the trend of values in the long term. Such variations can be smoothed to show the longer term trend by the use of running means (moving averages). This technique involves calculating the moving total and then the running mean for overlapping groups of years and then plotting this value at the middle year of each group. Each year therefore is allocated the average value of itself and its neighbouring years.

Figure 1.14 illustrates the construction of graph lines representing an annual value and two examples of running means. The table in Figure 1.14(a) shows how the running means have been calculated. For the 3-year running means the values have been divided into overlapping groups of 3. For each of these groups the moving total has been calculated. When this moving total is divided by 3 this gives the running mean which is allocated to the central year of the corresponding group of 3 years. Figure 1.14(a) also shows how the 5-year running mean has been calculated by dividing each moving total by 5.

When these running means are plotted on the graph it can be seen that the sudden inflow of Turkish labour in 1980 has been 'smoothed out' to some extent. The running means illustrate the longer term trend. On the graph it can also be seen that the 5-year running mean smooths the 1980 total more effectively than does the 3-year running mean. In fact, the 5-year running means show a slight decrease from 1979 to 1980. This is caused by the sharp decreases in Turkish inflow in 1981 and 1982 which are both included in the calculation for 1980. Thus one of the problems of using running means is that an upturn in a particular year can be represented as a downturn in the longer term.

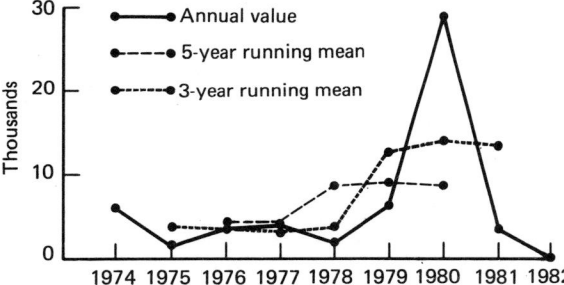

Fig. 1.14 Running means of the inflow of Turkish labour to West Germany (1974–82)

3-year running mean

1974	1975	1976	1977	1978	1979	1980	1981	1982	Moving total	Running Mean	Middle year
6.1	2.0	2.6	3.3	1.5	5.7	29.2	3.6	0.4			
6.1	2.0	2.6							10.7	3.6	1975
	2.0	2.6	3.3						7.9	2.6	1976
		2.6	3.3	1.5					7.4	2.5	1977
			3.3	1.5	5.7				10.5	3.5	1978
				1.5	5.7	29.2			36.4	12.1	1979
					5.7	29.2	3.6		38.5	12.8	1980
						29.2	3.6	0.4	33.2	11.1	1981

5-year running mean

6.1	2.0	2.6	3.3	1.5					15.5	3.1	1976
	2.0	2.6	3.3	1.5	5.7				15.1	3.0	1977
		2.6	3.3	1.5	5.7	29.2			42.3	8.5	1978
			3.3	1.5	5.7	29.2	3.6		43.3	8.7	1979
				1.5	5.7	29.2	3.6	0.4	40.1	8.1	1980

Fig. 1.14(a) Calculation of running means

Another problem with using the running mean is that, as the number of years in each group increases, there are more years at each end of the sequence that cannot have a value for the running mean. In Figure 1.14(a) no values for the 5-year running mean can be calculated for 1974, 1975, 1981 or 1982.

Combining two line graphs
In certain circumstances two separate line graphs can be combined to produce a new graph which illustrates the characteristics of both and which may in some cases make it possible for the reader to recognise and understand relationships in the data that are not easily recognised in the two separate line graphs. In geography the most common examples of this occur in 'time series' data where the horizontal axis is scaled in time values such as consecutive years or the months of the year and the vertical axis represents a single variable such as mean temperature, mean precipitation, or birth rate or death rate. In the case of a climatic graph based on mean monthly temperature and precipitation values the aim is to emphasise the relationships between mean monthly temperatures and mean monthly precipitation rather than between each of these separately and the months of the year. This can be best achieved by plotting mean monthly temperatures along the horizontal axis and mean monthly precipitation along the vertical axis. The months of the year can easily be labelled at each of the plotted points on the graph.

Figure 1.15(a) shows two climatic graphs of the traditional type relating to Location A and Location B. These graphs provide useful information about the relationship between mean monthly temperature and mean monthly

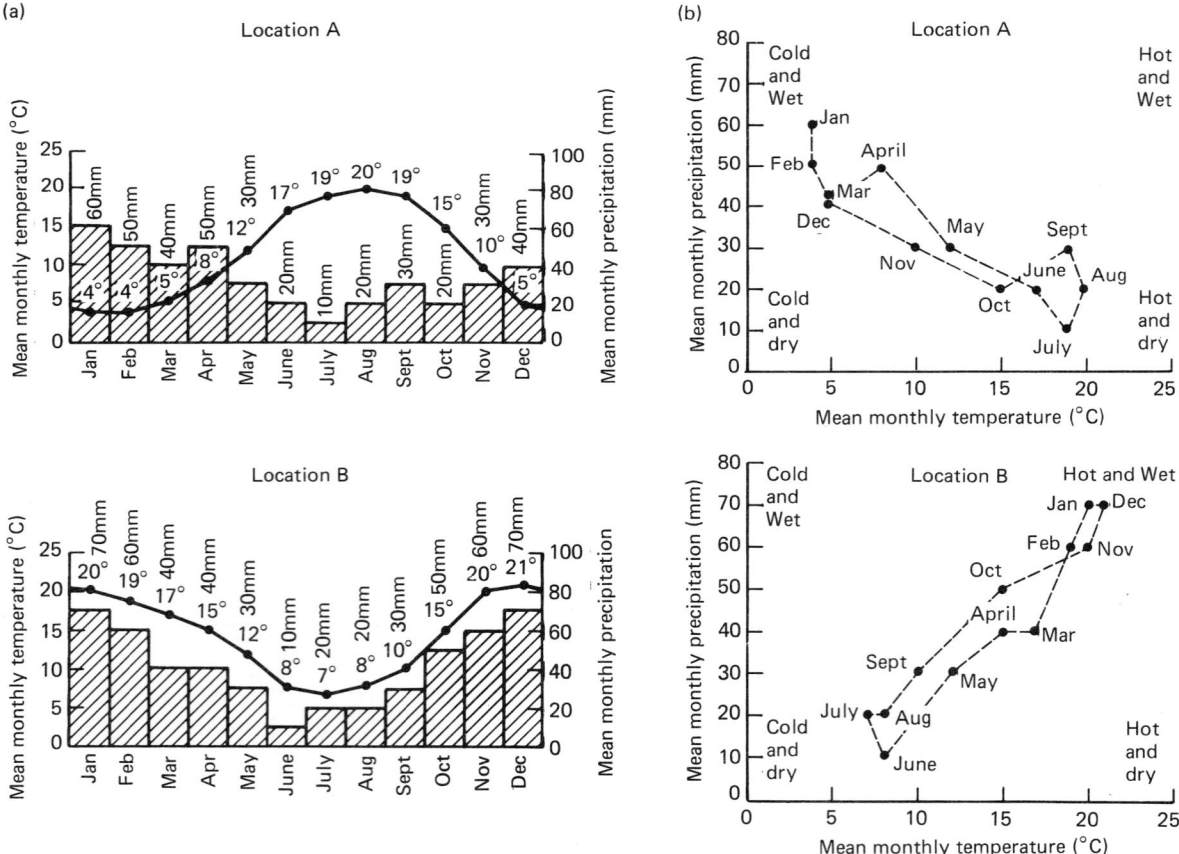

Fig. 1.15 Traditional climate graphs

precipitation through the year. Figure 1.15(b) shows the same information in the much more compact form of two line graphs (hythergraphs). These diagrams seem much easier to compare than the graphs in Figure 1.15(a). Also there is no need to construct a graph for each location. It is possible to plot 2, 3 or more sets of monthly climatic data for various locations on the same graph. Figure 1.16 shows very clearly the differences between the climate of a Mediterranean coast location and that of a location on the coast of Norway. It is possible to see at a glance the comparative ranges of both mean monthly temperature and mean monthly precipitation. Similarly it is easy to make detailed comparisons, such as the similarity between the Mediterranean coast climate in December and the Norwegian coast climate in May. It would just be possible perhaps to insert a third line graph into Figure 1.16 without causing excessive congestion.

Cumulative line graphs

Generally a cumulative line graph shows the number of items or values that are greater than or less than a certain level. One type of cumulative line graph, the ogive, has been explained earlier (page 5, Fig. 1.4). It was constructed from a simple histogram by adding the number of items in each class to the total number in the classes of lower value. There are other types of cumulative line graphs that differ slightly from an ogive.

An interesting example of a cumulative line graph is a graph showing *accumulated temperatures* (Fig. 1.17). This is constructed according to similar principles to those underlying the ogive. It is concerned with the relationships that may exist between temperature and plant growth. Figure 1.17 shows the mean daily temperature for each of the months of the year at Berlin and Scilly. As might be expected in view of their relative location, Berlin has higher

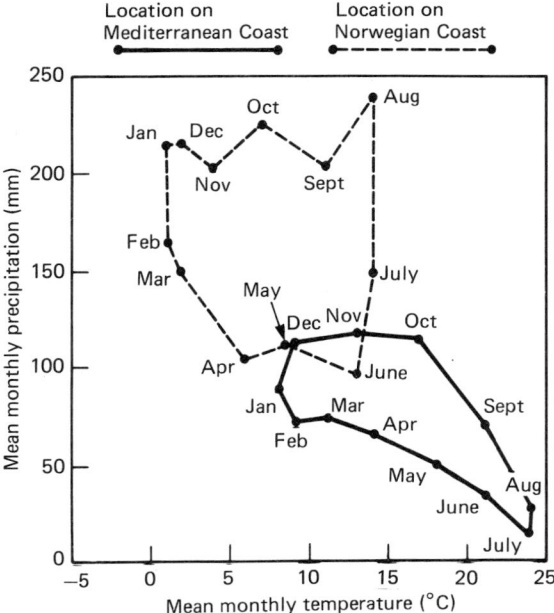

Fig. 1.16 A hythergraph

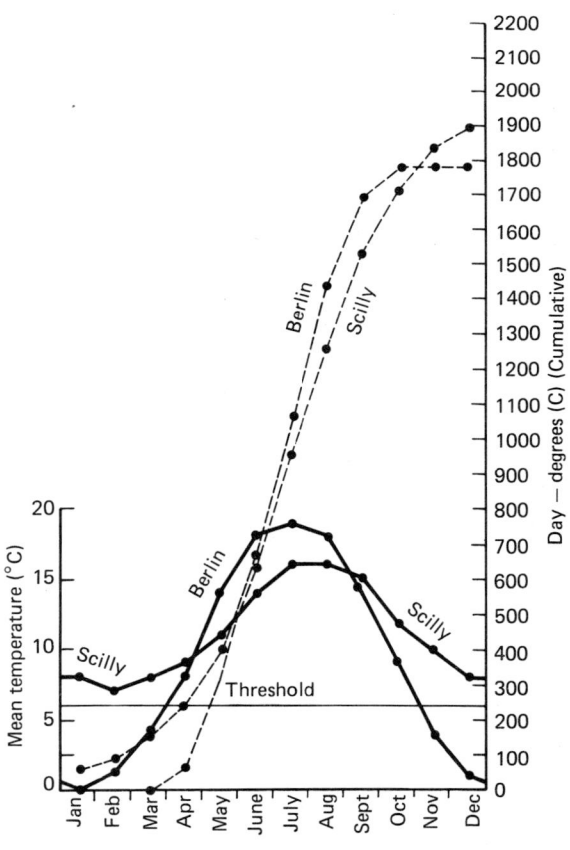

Fig. 1.17 A graph showing accumulated temperatures at Berlin and Scilly

16 PRACTICAL GEOGRAPHY: PRESENTATION AND ANALYSIS

Scilly	Mean daily temperature		Threshold			Number of days				day-degrees
January	8	–	6	=	2	31	×	2	=	62
February	7	–	6	=	1	28	×	1	=	28
March	8	–	6	=	2	31	×	2	=	62
April	9	–	6	=	3	30	×	3	=	90
May	11	–	6	=	5	31	×	5	=	155
June	14	–	6	=	8	30	×	8	=	240
July	16	–	6	=	10	31	×	10	=	310
August	16	–	6	=	10	31	×	10	=	310
September	15	–	6	=	9	30	×	9	=	270
October	12	–	6	=	6	31	×	6	=	186
November	10	–	6	=	4	30	×	4	=	120
December	8	–	6	=	2	31	×	2	=	62

Berlin	Mean daily temperature		Threshold			Number of days				Day-degrees
January	0	–	6	=	0	31	×	0	=	0
February	1	–	6	=	0	28	×	0	=	0
March	4	–	6	=	0	31	×	0	=	0
April	8	–	6	=	2	30	×	2	=	60
May	14	–	6	=	8	31	×	8	=	248
June	18	–	6	=	12	30	×	12	=	360
July	19	–	6	=	13	31	×	13	=	403
August	18	–	6	=	12	31	×	12	=	372
September	14	–	6	=	8	30	×	8	=	240
October	9	–	6	=	3	31	×	3	=	93
November	4	–	6	=	0	30	×	0	=	0
December	1	–	6	=	0	31	×	0	=	0

Accumulated temperatures for Scilly and Berlin

	Scilly	Berlin
January	62	0
February	90	0
March	152	0
April	242	60
May	397	308
June	637	668
July	947	1071
August	1257	1443
September	1527	1683
October	1713	1776
November	1833	1776
December	1895	1776

Fig. 1.17(a) Mean daily temperatures

temperatures than Scilly from May to August, but Scilly is warmer than Berlin from September through the winter to the following April. The length of the growing season for plants is clearly dependent on temperature, and a temperature of 6°C has been selected as the 'threshold' above which a great number of plants begin to grow. On the graph in Figure 1.17 a line labelled 'threshold' has been drawn aross the graph at 6°C. The accumulated temperatures *above* this threshold can be calculated from this graph, thus allowing the growing season at Berlin to be compared with that at Scilly.

This is done by calculating the 'day-degrees' for each month and plotting these values cumulatively with reference to the graph's right-hand axis. Each monthly total of day-degrees is calculated by subtracting 6°C, the threshold, from each monthly mean temperature and then multiplying by the number of days in the month. The calculations are set out in the tables in Figure 1.17(*a*), for both Scilly and Berlin. For example, in January Scilly's mean daily temperature is 8°C, from which 6°C, the threshold, is subtracted to leave 2°C, which is then multiplied by 31 (the number of days in January) to give a value of 62 day-degrees. This value has been plotted on the graph for Scilly in January. This kind of calculation is then carried out for the months of the year, for both Scilly

GRAPHS

and Berlin. (Refer to the tables in Figure 1.17(a).) Finally, the numbers of day-degrees at each place are summed cumulatively to give a table of accumulated temperatures which can then be plotted on the graph. The two locations can then be compared. Scilly appears to have a year-round growing season but Berlin's growng season appears to be restricted to seven months. Despite this, Berlin receives considerably more day-degrees than Scilly from May to August. Although Scilly enjoys more day-degrees of accumulated temperatures than Berlin for the year as a whole, Berlin's accumulated temperatures are greater than Scilly's from June to October. This can be seen from the cumulative curve on the graph.

It should be noted that both of the cumulative curves shown on the graph in Figure 1.17 have the S-shape that is generally typical of an ogive.

THE HYPSOMETRIC CURVE

A hypsometric curve is a cumulative line graph that relates the altitude of the land to its area. It shows the percentage of a total area (or sometimes the actual total area) that lies within a particular height range. The vertical axis of the graph shows the height of the land and the horizontal axis shows the area or the percentage of the total land area (Fig. 1.18(c)).

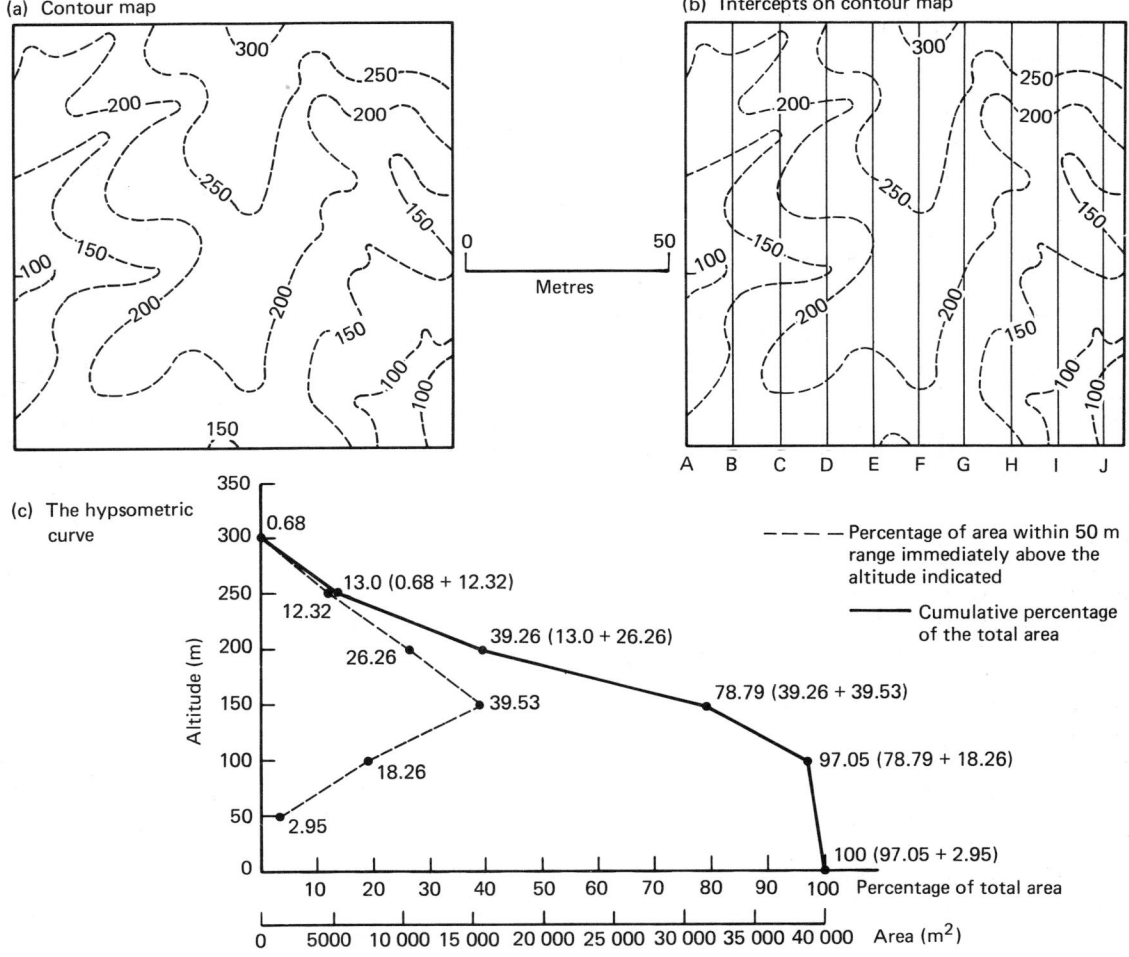

Fig. 1.18 The construction of a hypsometric curve

Table A   *Length of intercepts on contour map (mm)*

Altitude zone	A	B	C	D	E	F	G	H	I	J	Total
300 – 350 m	0	0	0	0	0	13	0	0	0	0	13
250 – 300 m	0	0	0	16	51	73	31	19	22	22	234
200 – 250 m	23	14	54	96	83	78	81	26	18	26	499
150 – 200 m	60	127	117	74	56	26	78	104	61	48	751
100 – 150 m	94	49	19	4	0	0	0	41	62	78	347
	13	0	0	0	0	0	0	0	27	16	56
Total	190	190	190	190	190	190	190	190	190	190	1900

Table B   *Percentages of area in altitude zones*

Altitude zone	Total length of intercepts	% of total area	Cumulative % of area
300 – 350 m	13	0.68	0.68
250 – 300 m	234	12.32	13.0
200 – 250 m	499	26.26	39.26
150 – 200 m	751	39.53	78.79
100 – 150 m	347	18.26	97.05
50 – 100 m	56	2.95	100.0

Fig. 1.18 (cont.) Data for the construction of a hyposymetric curve

The construction of a hypsometric curve

To construct a hypsometric curve it is necessary first to calculate the area of land that each altitude zone occupies. This makes it possible to relate the area of land surface to the various contour intervals. When this has been done it is comparatively easy to produce a hypsometric curve. One way to calculate areas on a contour map is to draw a series of parallel, equally spaced, lines across the map and then to measure along these lines the length of distances between the points where the line that has been drawn intersects with contour lines (Figure 1.18(*b*)). These distances are termed intercepts. These total lengths are likely to give a reasonable estimate of the comparative areas of land occupied by the various altitude zones. The sample parallel lines, (A to J) are indicated in Figure 1.18(*b*). The measurement process needs to be done with great care, because the lengths of the contour intercepts for each line must be equal to the total length of the line. It is a good idea to measure the intercepts in millimetres so that there need be no decimal values.

Table A shows the lengths of these contours intercept along each of the lines A to J. Table B shows how the lengths of the intercepts of the various altitude zones can be expressed as percentages of the grand total (i.e. 1900) and therefore as percentages of the total area of the map. By adding these percentages consecutively we determine the cumulative areas occupied by the various altitude zones. The resultant percentage hypsometric curve is shown in Figure 1.18(*c*).

Figure 1.18 makes it clear that the hypsometric curve is simply the cumulative version of the graph indicated by the dashed lines. The hypsometric curve in Figure 1.18(*c*) seems a little peculiar in that it is suddenly 'chopped off' at the right-hand end. This is because there are so few altitude values below 100 metres on the original map. It should be noted that, in Tables A and B and the graph, the percentage values have been cumulated from the highest values downwards towards the lower values. This is usual for a hypsometric curve. It indicates the percentage of the area that has an altitude of 'more than' a given value rather than 'less than'.

Finally, it has been possible to include in Figure 1.18(*c*) some information about the height of actual areas of ground in addition to percentages of the area. The study area (Fig.1.18(*a*)) has an area of 200 metres by 200 metres, i.e. 40 000 square metres. It has therefore been possible to insert an areal scale along the horizontal axis of Figure 1.18(*c*) with 40 000 square metres corresponding to 100%.

GRAPHS

THE LORENZ CURVE

The Lorenz curve illustrates the degree of unevenness in a geographical distribution. It can be used to show irregularities in the distribution of population, incomes or the production of various commodities.

The construction of a Lorenz Curve

It is drawn on a square graph with the X and Y axes having similar scales that are usually expressed in percentages. On the graph a perfectly even distribution is shown by a straight line running diagonally across it. On the graph in Figure 1.19, for example, the diagonal line links together the points representing equal percentages of the area and the population (i.e. equal densities of population). The extent to which the Lorenz curve deviates from this diagonal line shows the extent to which the actual distribution differs from a perfectly regular distribution. In Figure 1.19 there is a point on the Lorenz curve where, instead of about 50% of the area containing about 50% of the population, as in a regular distribution, 50% of the area contains over 90% of the population, a disparity of over 40%. The table in Figure 1.19 illustrates this. The cumulative percentage of the total area from the Netherlands to France inclusive is 49.51 (column (iv)) but, for the same area, the cumulative percentage of the population is 90.69 (column (vii)).

Figure 1.19 has been constructed in the following way. Information has been collected about the areas and population totals of a sample of 10 countries in NW Europe. The aim is to analyse the pattern of national population densities in NW Europe. First of all the countries are listed in descending order of population density (column (i)). On the graph it is necessary

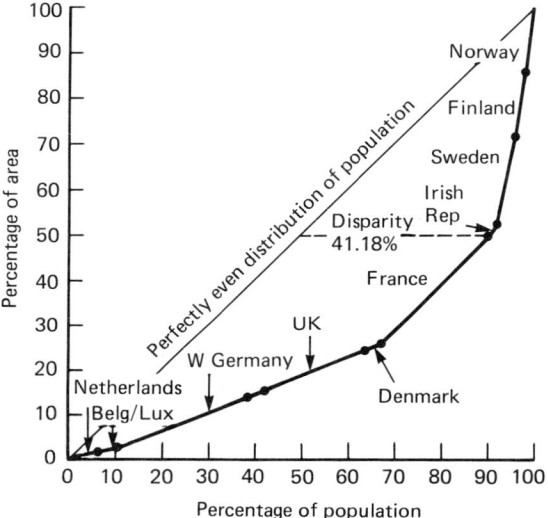

(a)   Percentage of population

	(i) Population density per sq km	(ii) Area (thousands of sq km)	(iii) % of total area	(iv) Cumulative % of total area	(v) Population (millions)	(vi) % of total population	(vii) Cumulative % of total population
Netherlands	350	41	1.75	1.75	14.3	6.43	6.43
Belgium/Lux	309	33	1.41	3.16	10.2	4.58	11.01
West Germany	247	249	10.65	13.81	61.5	27.64	38.65
United Kingdom	230	245	10.47	24.28	56.5	25.39	64.04
Denmark	119	43	1.84	26.12	5.1	2.29	66.33
France	99	547	23.39	49.51	54.2	24.36	90.69
Irish Republic	50	70	2.99	52.50	3.5	1.57	92.26
Sweden	19	450	19.24	71.74	8.3	3.73	95.99
Finland	14	337	14.41	86.15	4.8	2.16	98.15
Norway	13	324	13.85	100.00	4.1	1.85	100.00
		2339	100.00		222.5	100.00	

Fig. 1.19 A Lorenz curve based upon variations in population density in NW Europe

to plot the countries in this order from the origin of the graph. In column (*ii*) the areas of the countries are listed and these are summed to give the total area of the whole sample of countries. It is easy now to calculate the percentage of the whole area that is occupied by each country (column (*iii*) and this column must total 100.

In column (*iv*) the values of column (*iii*) are summed cumulatively, the final value being 100, as in the column (*iii*) total. Next, exactly the same procedure is carried out for the population totals of the countries. In column (*v*) the totals are summed to give the total population of 222.5 millions. Then, in column (*vi*), each country's percentage of the total population is listed, and finally these values are summed cumulatively in column (*vii*). The graph is now constructed using the two columns of cumulative percentages (*iv*) and (*vii*)), column (*iv*) being applied to the Y axis and column (*vii*) to the X axis.

It is easy to make a comparison between the distributions illustrated in Figures 1.19 and 1.20. Wheat production (Fig.1.20) is clearly distributed more evenly over Europe than population is in NW Europe. The Lorenz curve in Figure 1.20 shows that 50% of the wheat production is spread over more than 30% of the area. In Figure 1.19, 50% of the population is concentrated in less than 20% of the area. Similarly, 80% of the wheat production is spread over about 60% of the area whereas 80% of the population is concentrated in about 40% of the area. It would be possible of course to plot several Lorenz curves on the same graph and

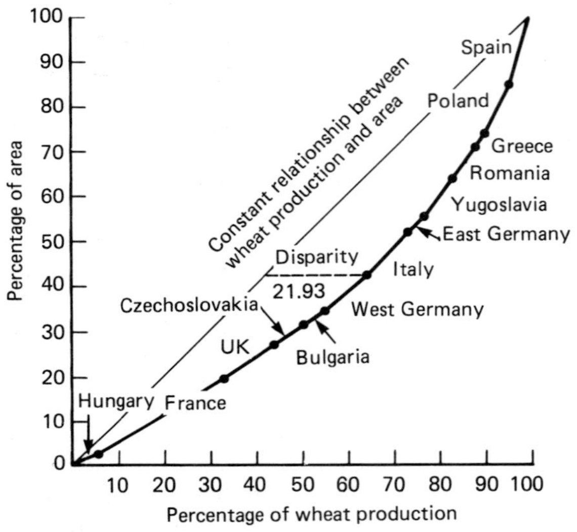

(a)

	(i) Tonnes per sq km	(ii) Area (thousands of sq km)	(iii) % of total area	(iv) Cumulative % of total area	(v) Wheat production (millions of tonnes)	(vi) % of total wheat production	(vii) Cumulative % of total wheat production
Hungary	59.14	93	2.88	2.88	5.50	5.96	5.96
France	45.34	547	16.96	19.84	24.80	26.86	32.82
United Kingdom	43.27	245	7.59	27.43	10.60	11.48	44.30
Czechoslovakia	41.41	128	3.97	31.40	5.30	5.74	50.04
Bulgaria	39.64	111	3.44	34.84	4.40	4.77	54.81
West Germany	35.89	249	7.72	42.56	8.94	9.68	64.49
Italy	27.91	301	9.33	51.89	8.40	9.10	73.59
East Germany	27.78	108	3.35	55.24	3.00	3.25	76.84
Yugoslavia	21.56	256	7.94	63.18	5.52	5.98	82.82
Romania	21.01	238	7.38	70.56	5.00	5.42	88.24
Greece	15.80	132	4.09	74.65	2.09	2.26	90.50
Poland	14.15	313	9.70	84.35	4.43	4.80	95.30
Spain	8.6	505	15.65	100.00	4.35	4.71	100.01
		3226	100.00		92.33	100.01	

Fig. 1.20 *A Lorenz curve based upon variations in the intensiveness of wheat production in Europe*

GRAPHS

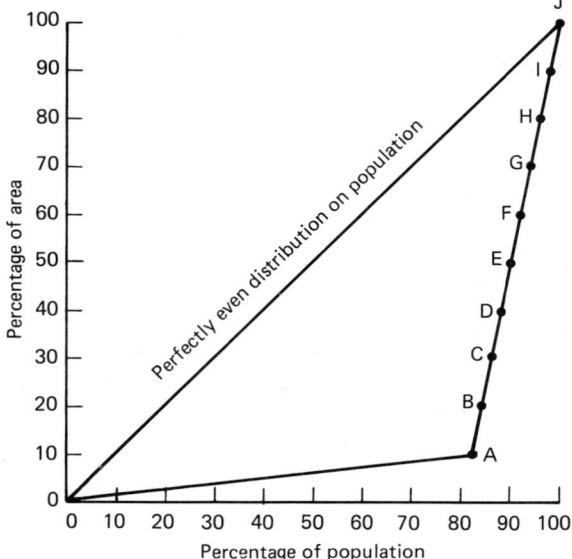

Fig. 1.21 An extreme example of concentration of population

thus make comparisons rather easier to see.

Figure 1.21 illustrates an imaginary population distribution that is extremely concentrated. Over 80% of the population is concentrated in 10% of the area. Comparing the three graphs (Figs. 1.19, 1.20 and 1.21) it can be seen that the degree of concentration of the distribution is indicated by the size of the area between the diagonal line and the Lorenz curve. Figure 1.20 has the smallest such area and also has the lowest degree of concentration. Figure 1.21 has the largest area and also the highest degree of concentration.

Theoretically of course we could construct a Lorenz curve to show a situation in which 99% of the population is concentrated into 1% of the area. In this case we would be nearing the maximum possible size for the area between the Lorenz curve and the diagonal line.

It is possible to describe the degree of regularity or concentration of a distribution by calculating a *similarity index*.

$$\text{Similarity Index} = 1 - \frac{\Sigma \text{ differences (\%s)}}{100}$$

To calculate this index for the data of Figure 1.19 we first of all sum the differences between the successive values in column (*iii*) (% of total area) and their corresponding values in column (*vi*) (% of total population). The sum of these differences is divided by 100 and the resultant value is subtracted from 1. Clearly, if there were no difference at all between the percentages for total area and total population the similarity index would be 1 (i.e. 1 − 0). The distribution shown in Figure 1.19 is extremely irregular and its similarity index has a low value of only 0.18. This is not really surprising because the percentage differences between area and population are over 10 in the cases of West Germany and the United Kingdom (with high population totals in relation to their area) and also for Sweden, Finland and Norway (with very low population totals and quite large areas).

The distribution shown in Figure 1.20 is much more regular and it has a similarity index of 0.56. Only Spain and France have large percentage differences, France having a very large wheat production in relation to its area and Spain having just the opposite, a small wheat production in relation to its area. The practically impossible distribution shown in Figure 1.21 has a similarity index of −1.44. One of the percentage differences rises to 72. Perhaps a dissimilarity index would be more appropriate in this case!

GRAPHS WITH LOGARITHMIC SCALES

Logarithms

A logarithm is the *power* to which it is necessary to raise the number to reach a particular number. For example, the logarithm of 10 is 1, but the logarithm of 100 is 2, because, to reach 100, 10 has to be squared. 10 × 10 × 10 makes 1000, so the logarithm of 1000 is 3. There are also logarithms for numbers less than 10 but, in these cases, they are negative. For example, the logarithm of 1 is −1 and the logarithm of 0.1 is −2.

However it is not necessary to know very much about logarithms before using logarithmic graphs because there is no need to plot the actual logarithms of the numbers on such graphs. Instead graph paper with logarithmic scales on both axes is available and this is as easy to use as plain graph paper (Fig.1.22(*a*)). The bold lines on this graph paper are labelled successively 1, 2, 3, 4, 5, 6, 7, 8, 9 and then 1 again. The intervals between these values gradually decrease in width. In Figure 1.22(a) the axes have been labelled from 1 to 10, but in other cases it is quite permissible to add zeros to change the range to either 10 to 100 or 100 to 1000 or even 0.1 to 1. It is not possible to have a value for zero on a logarithmic graph because zero has no logarithm. The possible range of values on the axes of the graph can be increased greatly by including a number of successive cycles of values on each axis. In Figure 1.23(b), for example the Y axis has a range extending from 10 thousands to 10 millions in three successive cycles of 10 to 100 and 100 to 1000 (in thousands), followed by 1 to 10 in millions. Along the X axis the first cycle is 1 to 10, the next is 10 to 100 and the last is 100 to 1000.

Logarithmic graphs and semi-logarithmic graphs

There are two kinds of graph with a logarithmic scale. The graph in Figure 1.22(a)(i) is termed a 'logarithmic graph' and has a logarithmic scale on both axes. A 'semi-logarithmic graph' (Fig. 1.24) has a logarithmic scale on only one of its two axes. These two kinds of graph serve quite different purposes.

Figure 1.22(a) shows the difference in the appearance between a graph with ordinary arithmetic scales along its axes ((a)(ii)) and a graph with logarithmic scales ((a)(i)). Although the graphs look so different they are displaying exactly the same information. The chief difference between them is the magnification of the scale of the intervals between the plotted lines in the lower left-hand corner of (a)(i) and the sharp reduction in the scale in the top right-hand corner of (a)(i). This has resulted in the substitution of curves in (a)(i) for the straight lines in (a)(ii). This suggests that (a)(i) might be suitable for plotting a graph in a case where a very large proportion of points fell between line C and the origin. From line C outwards the width of the plotted zones are greater in (a)(ii) than in (a)(i). This suggests that more extreme values than those shown would fit more easily into (a)(i) than (a)(ii). The main point however is that straight graph lines become curved when plotted on a logarithmic graph.

In Figure 1.22(b) the opposite occurs. In (b)(i) a logarithmic graph has parallel straight

Fig. 1.22 Logarithmic graphs (a)(i) and (b)(i) and their corresponding arithmetic graphs (a)(ii) and (b)(ii)

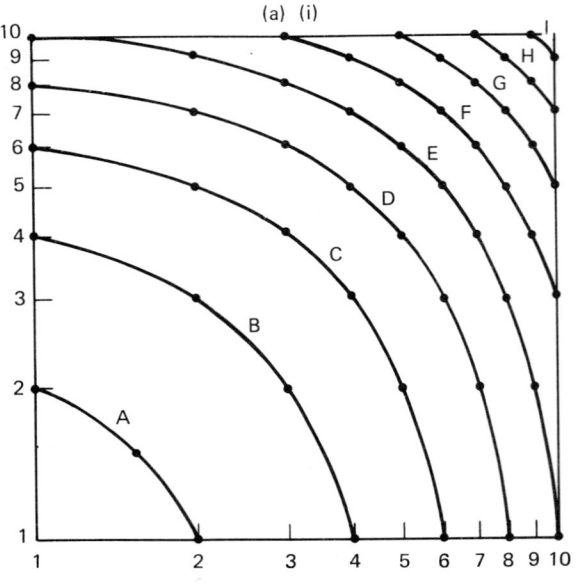

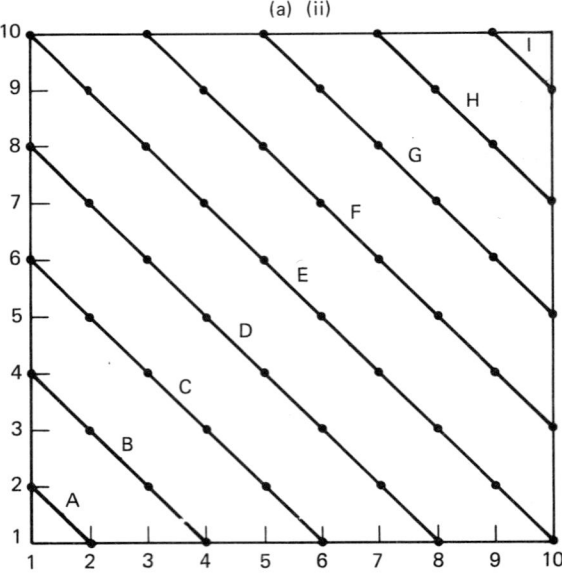

GRAPHS

lines. When this is plotted in relation to arithmetic scales ((b)(ii)) the lines become curved and drawn tightly in towards the origin of the graph. Thus, even when they are displaying exactly the same information, there is a great difference in

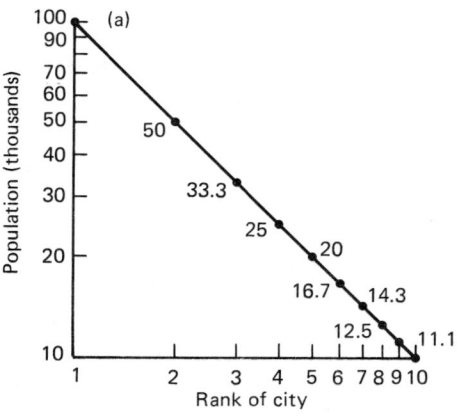

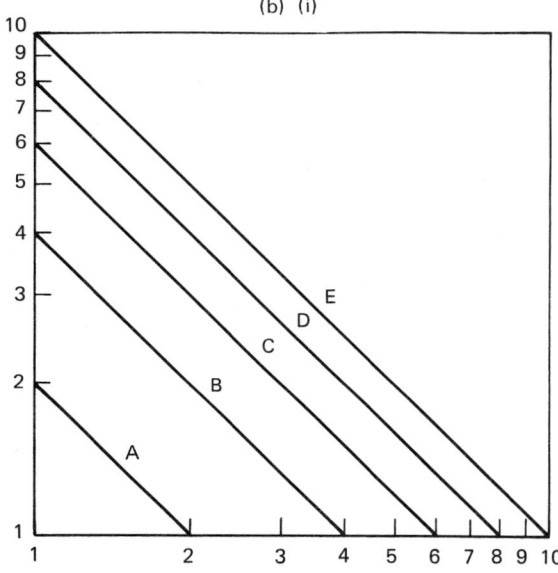

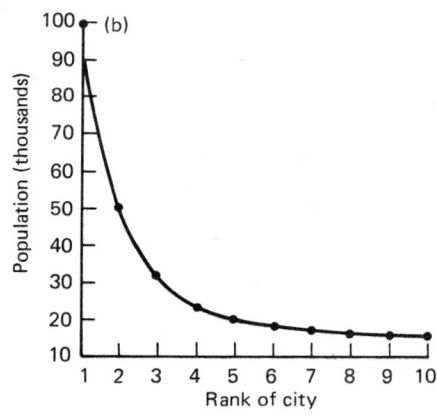

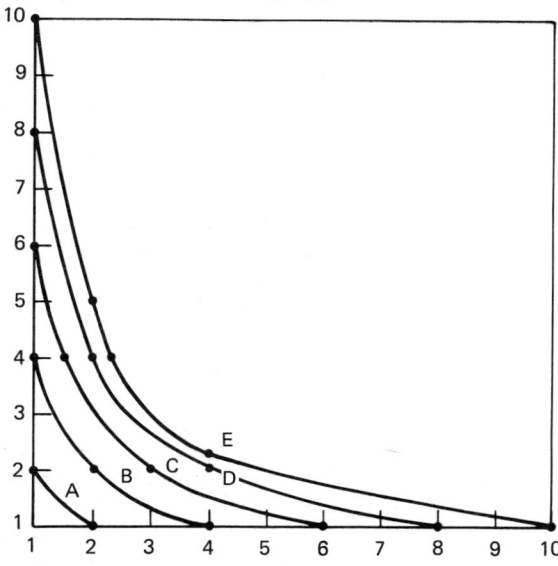

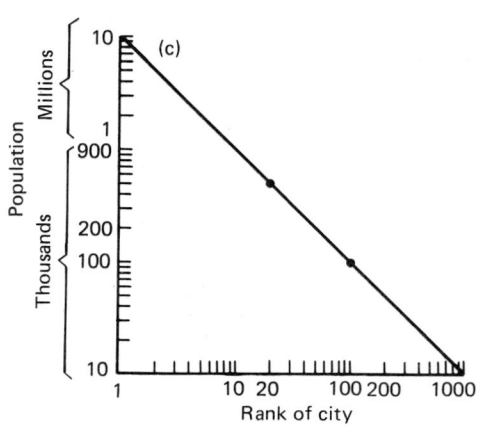

Fig. 1.23 The rank-size rule

appearance between graphs with arithmetic scales and those with logarithmic scales.

Many of the relationships that have been identified in both physical and human geography can be illustrated by the use of logarithmic graphs. A good example is the rank-size rule which is illustrated in Figure 1.23. In Figure 1.23(a) the diagonal line shows the population totals of a number of cities. The largest city of all has a population of 100 000. The second largest has 50 000 (i.e. 100 000 divided by 2), the third largest 33 300 (100 000 divided by 3) and so on. The logarithmic graph illustrates this rank-size relationship particularly clearly. The accompanying arithmetic graph, showing exactly the same information is not so easy to understand. Figure 1.23(b) shows rank-size relationships on a very large scale with three logarithmic cycles on each axis. The largest city has a population of 10 million and this graph makes it possible to estimate, according to the rank-size rule, the population totals of 999 cities. The thousandth city (X axis) has a population total of 10 000 (Y axis). Logarithmic **graphs** are also valuable in the study of **hydrology**. Regularities often exist in the relationships between, for example, river discharge and the area of the river's catchment, and also between the relief and the drainage density of a drainage basin.

In semi-logarithmic graphs it is also common to be able to identify related geographical variables. Figure 1.24 is a diagram of a drainage network in which the streams have been ordered according to Strahler's system. When the details are plotted on the semi-logarithmic graph the relationship between stream order and stream number becomes evident. There is one 4th order stream, two 3rd order streams, four 2nd order and eight 1st order streams. This means that between successive orders the stream numbers change by a constant factor of 2. They simply double each time.

Semi-logarithmic graphs are perhaps most widely used to interpret changes through time. These changes could relate to population growth or decline, increases or decreases in output of various commodities or imports or exports.

Figure 1.25 illustrates the difference between a normal arithmetic graph and a semi-logarithmic graph in their ability to portray rates of change. In Figure 1.25 (a) the straight diagonal line shows how the population total has increased

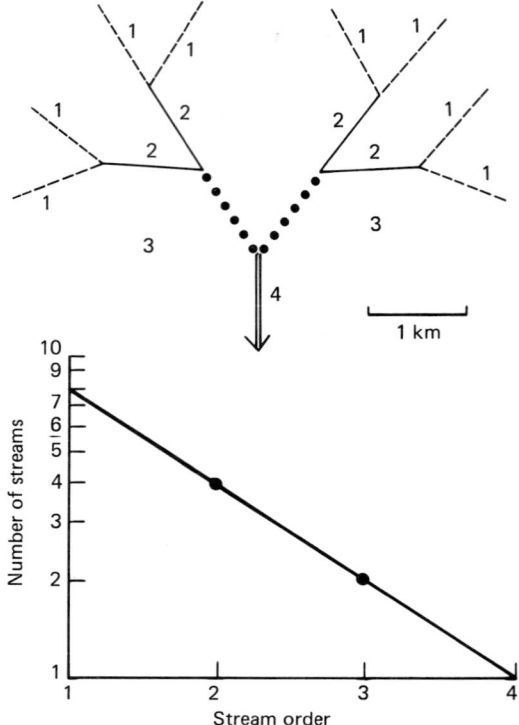

Fig. 1.24 Semi-logarithmic graph of stream order and number

between 1981 and 1986. It is a perfectly straight line and it shows that the population total has increased by 10 thousand in each year. This, however, is *not a constant rate* of population increase. In 1981–82 the population increased from 10 thousand to 20 thousand (an increase of 10 thousand from a starting point of 10 thousand in 1981). This is therefore an increase of 100%. The population has doubled. From 1982 to 1983 the population increased from 20 thousand to 30 thousand, that is, an increase of 10 thousand from a starting point of 20 thousand (i.e. a 50% increase). To maintain the 100% rate of increase in 1982–83 the population would have had to double itself to 40 thousand.

The situation becomes very much clearer when we refer to the semi-logarithmic graph (b). The population totals plotted here are exactly the same as those on the arithmetic graph. However, the constant rate of increase, taking the population total from 20 thousand in 1982 to 40 thousand in 1983, is shown as a straight line. Figure 1.25(b) seems very much more useful than Figure 1.25(a) not only because it can accommodate a greater range of population totals on the vertical axis but also because

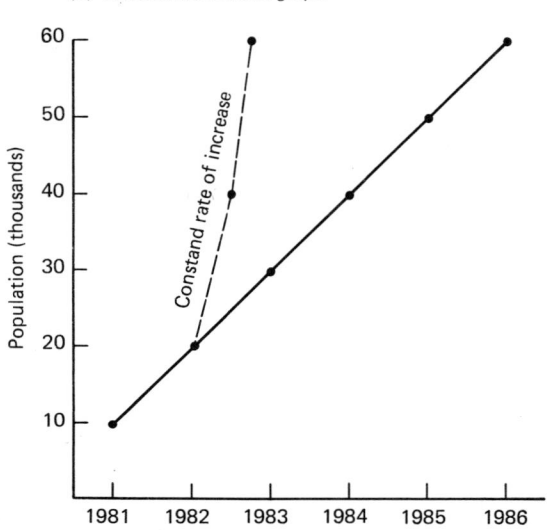

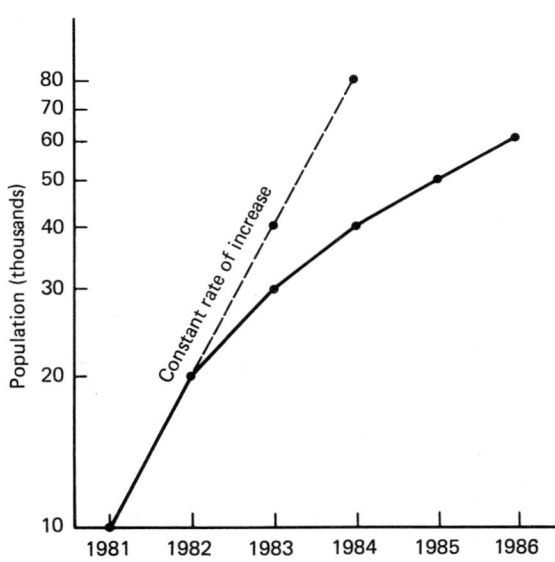

Fig. 1.25 Rates of change on normal arithmetic and semi-logarithmic graphs

it could help in making forecasts of the future population of the area based on assumptions concerning the likely rate of increase. The curving line on the semi-logarithmic graph is the exact equivalent of the straight line in Figure 1.25(a). This relationship makes it quite clear that the straight line in Figure 1.25(a) represents a decreasing rate of population increase.

Figure 1.26 summarises some of the detailed characteristics of semi-logarithmic graphs. First of all, in all such graphs the rate of increase or decrease of the variable is indicated by the gradient of the line. If two lines on a semi-logarithmic graph have exactly the same gradient, wherever they are situated on the graph, they will represent exactly the same rate of increase or decrease. This is indicated in Figure 1.26(a). Figure 1.26(b) illustrates another basic principle. This is that the gradient of the graph line indicates the rate of change of the variable. A steeply sloping line on the graph always indicates a faster rate of increase (or decrease) than a more gradually sloping one. Another principle is illustrated in Figure 1.26(c). The graph line here slopes less steeply as it reaches the higher values of the variable (population), but in this case the population is expanding at a constant rate of 10 thousand per year. However, each of these increases of 10 thousand constitutes a decreasing proportion of the total population. Hence the percentage increase gradually decreases. In 1981 – 82 the percentage increase was 100% (from 10 to 20 thousand) but in 1985 – 86 it was only 20% (from 50 to 60 thousand). Finally, Figure 1.26(d) shows three graph lines each with a different slope and therefore having different rates of increase. In all three cases however the population has increased by exactly the same amount and the same percentage. In all cases the 1986 population total is 80 thousand and in all cases it was once 10 thousand. As can be deduced from the gradients of the curves however, the increases have taken place at greatly varying speeds.

Figure 1.27 shows information about the development of the population of Japan since the 1920s. This information has been plotted on a semi-logarithmic graph, so we are able to deduce information relating to the percentage changes in successive years. Clearly the total population of Japan, apart from a pause in the early 1940s, has increased steadily at a rate of

26 PRACTICAL GEOGRAPHY: PRESENTATION AND ANALYSIS

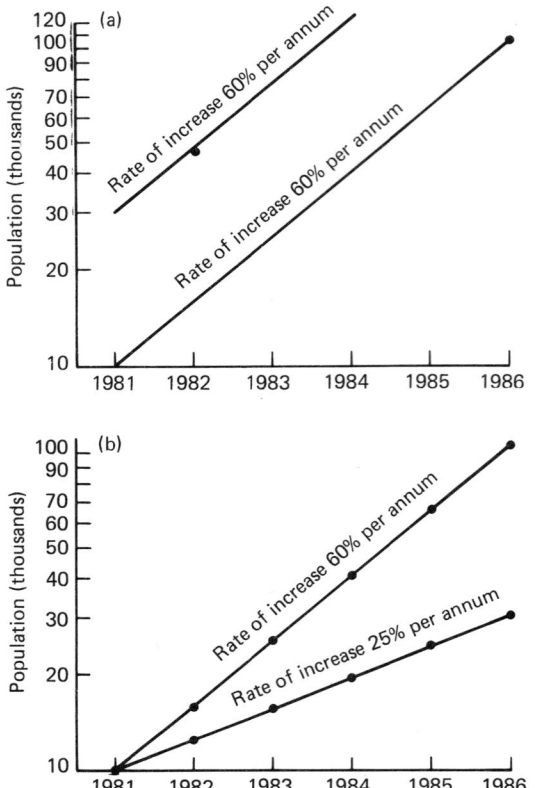
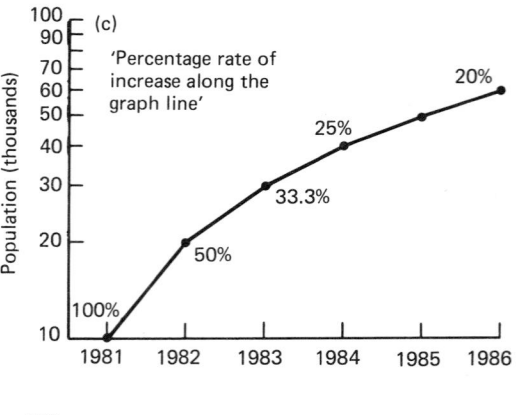
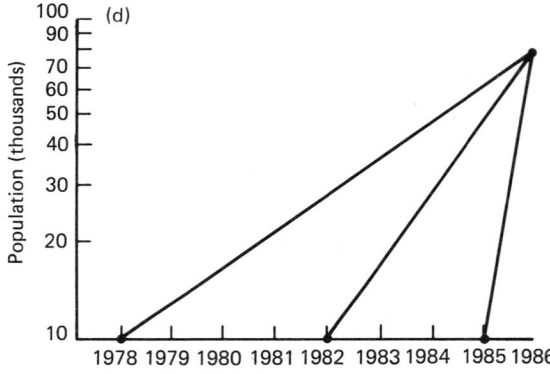

Fig. 1.26 Rates of change in semi-logarithmic graphs

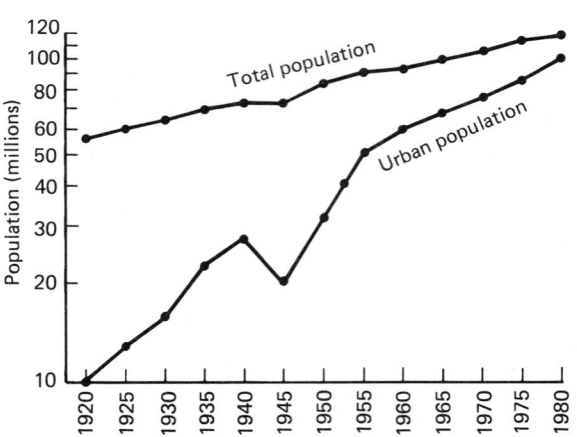

Fig. 1.27 Population growth in Japan (1920–80)

about 6% or 8% per year. The urban population has shown greater fluctuations. There was a decrease of about 27% in 1940 – 45 but this was followed by a period of rapid increase. In both 1945 – 50 and 1950 – 55 there were increases of 50% or more in the urban population. Since 1955 the urban population has been increasing by about 10% in each five-year period, which is rather faster than the Japanese population as a whole.

1.4 Scatter diagrams (scattergraphs)

A scatter diagram is a graph which shows the relationship between two or more variables by a distribution of dots. If there are two variables

GRAPHS

one of these is allocated to the X axis and the other to the Y axis. In a scatter graph there is not necessarily a dependent and an independent variable in the sense of one of the variables influencing the size of the other. However, some scattergraphs, by the combination of purely descriptive variables are capable of producing new knowledge. They can function as 'ready-reckoners'. In Figure 1.28, for example, the basic grid lines of the graph show the population totals and the areas of a selection of countries in the form of a scattergraph. However, by combining the population values with the area values it has been possible to produce a scale of population density per km². We can read off from the graph the area of the country, its total population and also its population density. We can see at a glance for example that Morocco's population density is just under 50 per km² and that Italy's is just under 200 per km². In this graph the general rule is that population densities increase towards the bottom right-hand corner where high total populations coincide with low areas.

Figure 1.29 is a similar graph. This time the two variables are the birth rate and the death rate. From these it is easy to construct lines representing the natural increase of population. The fastest growing countries are those such as Mexico and Egypt where very high birth rates occur in conjunction with quite low death rates. But we can also see countries with high death rates, such as Ethiopia and Gambia that still have high rates of natural increase simply because their birth rates are so high. Scattergraphs such as these are useful because they produce new knowledge.

Scattergraphs can also be used to deal with three variables instead of two. Three axes are required, so a triangular graph is used. Figure 1.30(*a*) shows how such a graph is set out. The values of the three variables are recorded successively around the perimeter of the triangle, each variable being measured in percentages.

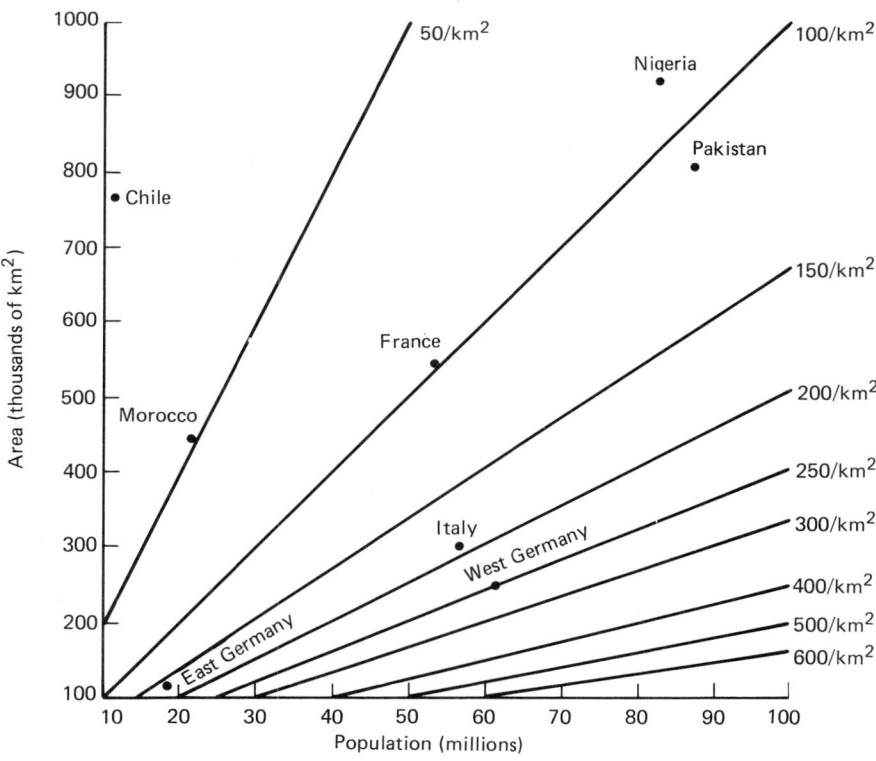

Fig. 1.28 Relationships between population and area

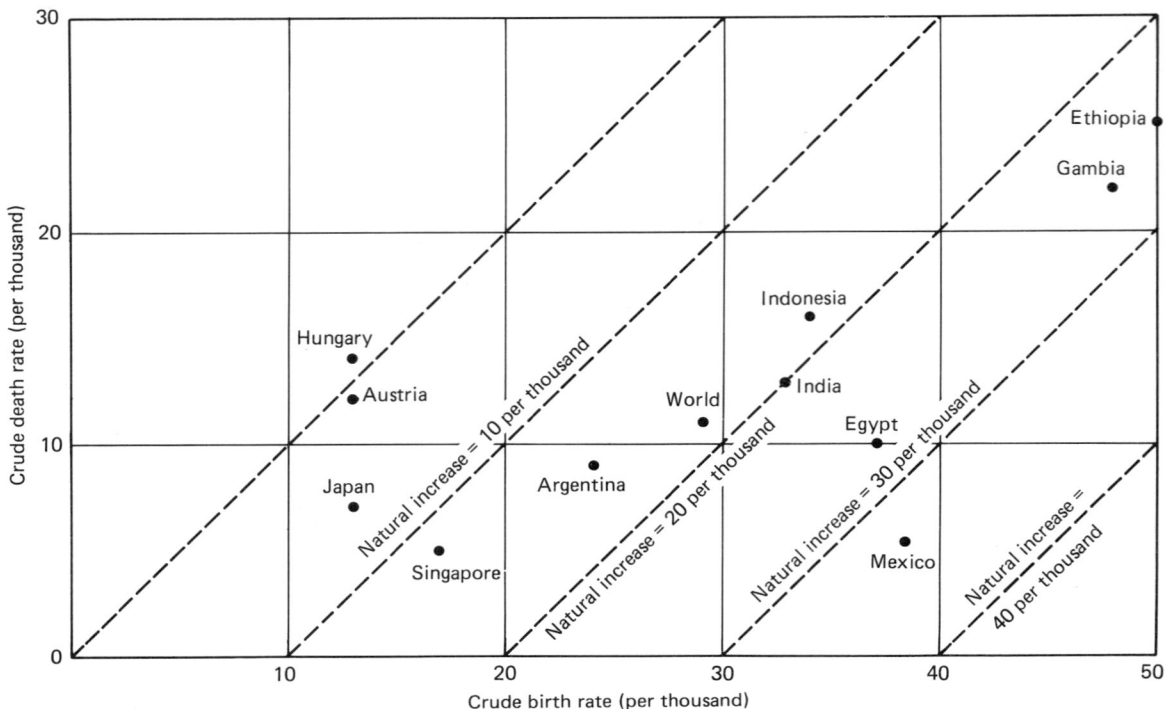

Fig. 1.29 Relationships between birth rate and death rate

Care has to be taken, when plotting values, to remember that the grid lines run at different angles. The three sets of consecutive percentage values from 0 to 100, follow each other consecutively round the perimeter of the graph, but the lines indicating the various percentage values run at different angles for each of the three sets.

Figure 1.30(b) shows some of the relationships that are built into a triangular graph and which are useful in interpreting the distribution of points on the scattergraph. It should be remembered that at any point on the graph the percentages of the three variables sum to a total of 100. After a little practice interpretation is quite easy. Variable A increases in value towards the bottom right-hand corner. Here therefore there is a zone where the values of variable A are greater than the sum of variables B and C. Moving towards the left, along variable B's axis, we see that B dominates the bottom left-hand corner. Similarly, variable C dominates the upper part of the graph. A triangle in the centre of the graph shows an area where no variable has a value greater than the sum of the other two variables. In the very centre of the graph there is a point where all the variables have equal values of $33\frac{1}{3}\%$.

The graph therefore is quite easy to interpret. The points near the centre have fairly equal percentages of the three variables. Points near the top are dominated by variable C; those near the bottom right-hand corner by variable A and those near the bottom left-hand corner by variable B. Figure 1.30(c) shows the distribution on the triangular graph of a number of European countries and the USA and Japan. As the countries have developed economically there has been a tendency for employment in agriculture to decline and for employment in services to increase. Greece and Portugal still have quite high percentages for agriculture (about 30%) but they only have about 40% for services. The USA, in contrast, has less than 5% employed in agriculture and almost 70% employed in services.

GRAPHS

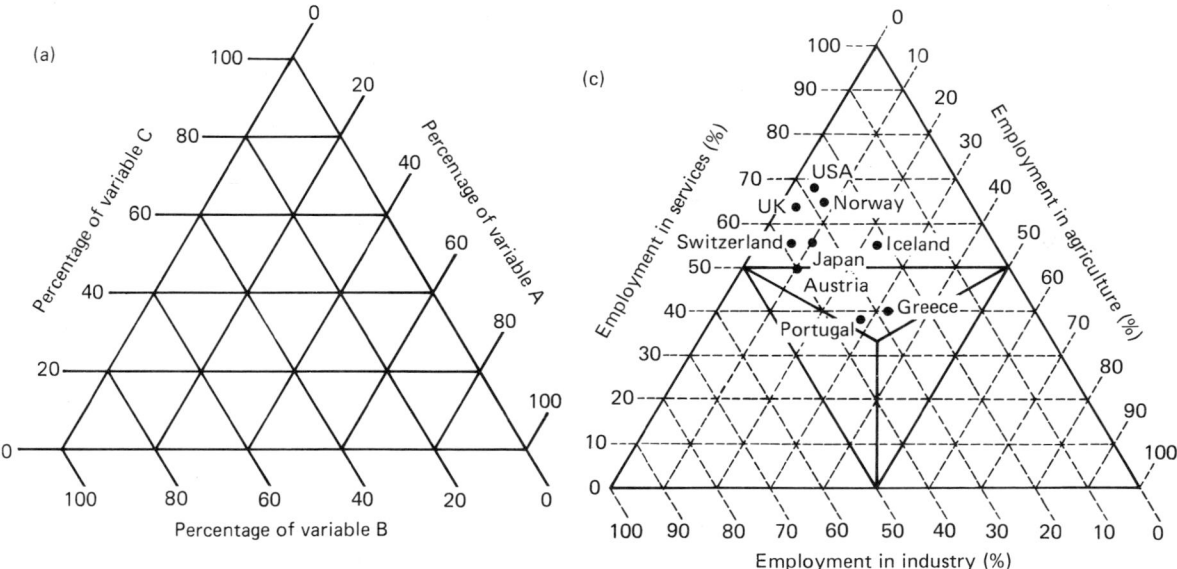

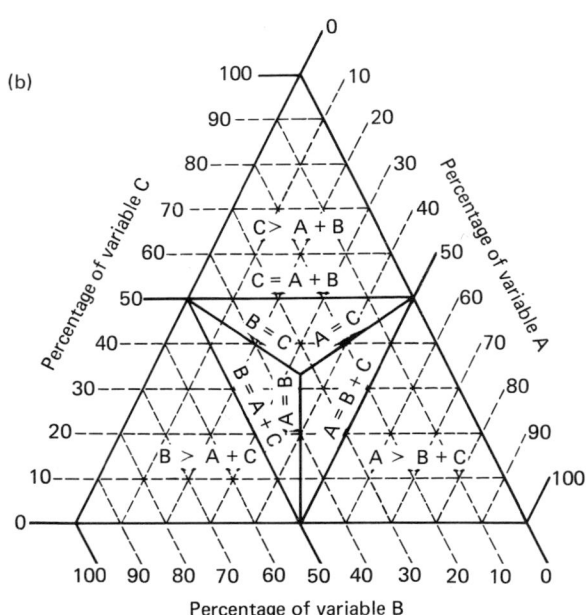

Fig. 1.30 The triangular graph

Other types of scattergraph have an independent variable scaled on the X axis and a dependent variable on the Y axis. Very frequently in geography we identify the nature of the relationship between the independent variable and the dependent variable. Figure 1.31 shows diagammatically some of the general relationships that can exist between dependent and independent variables. In Graph A the points form a line, so the variables are said to be linearly correlated. Even if they did not form a line but were confined in a narrow belt they would still be said to be linearly correlated. In Graph A also the variables are inversely (negatively) correlated. The large values of Y correspond to the small values of X. Graph B's variables are also linearly correlated but this time the correlation is positive. The variables in Graphs C and D are certainly not linearly correlated. Graph C represents an inverse (negative) correlation like Graph A. In Graph D the variables are positively correlated as in Graph B. In Graphs C and D the strength of the correlation is much weaker than in Graphs A and B. Finally in Graph E we have a case in which there is no recognisable correlation between the variables.

PRACTICAL GEOGRAPHY: PRESENTATION AND ANALYSIS

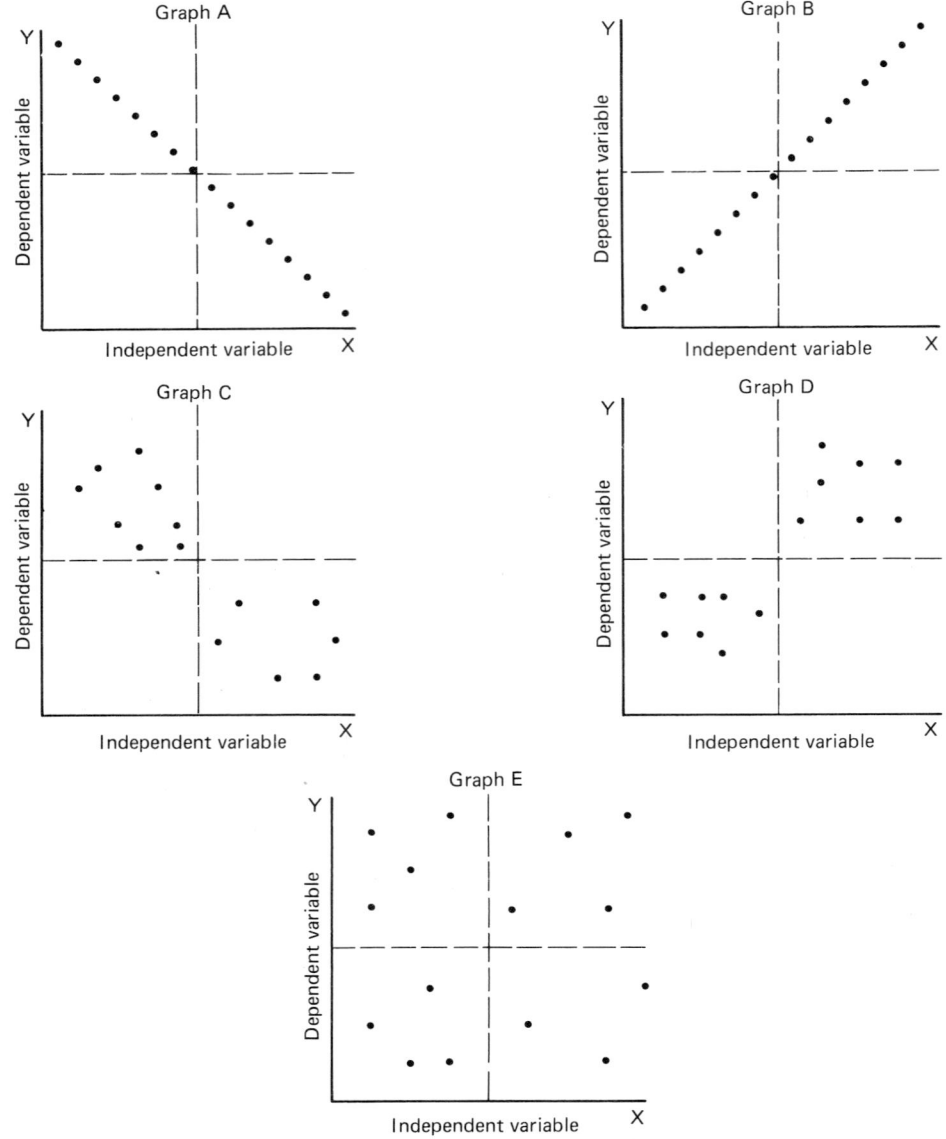

Fig. 1.31 Possible relationships between dependent and independent variables

1.5 Best-fit line

It is not easy to identify a general trend in the pattern of dots that exists in a scattergraph. It is much more satisfactory to deal with a line instead. It is quite simple to transform a dot pattern into a line that summarises its characteristics. This line is called a *best-fit line* or a *regression line*.

Constructing a best-fit line

Figure 1.32 shows a pattern of 16 dots and the best-fit line that summarises this pattern. Since X is the independent variable and Y is the dependent variable the best-fit line represents the regression of Y on X (i.e. how the value of Y changes in response to changes in X's value). The regression line for X on Y would be slightly different. The procedure is as follows:

(i)   Calculate the mean of the Y values shown by the dots;

GRAPHS

(ii) Calculate the mean of the X values shown by the dots;
(iii) Mark the point where the co-ordinates of these two means cross. The regression line must pass through this point;
(iv) Draw a line parallel to the Y axis through a point where the co-ordinates of the two means cross;
(v) Calculate and mark the mean of the points to the left of this line (M1 in Figure 1.32) and then the mean of the points to the right of it (M2 in Figure 1.32);
(vi) The regression line is now drawn so as to pass through the points representing the three means.

This line summarises the characteristics of the scattergraph.

It is possible to describe the characteristics of a regression line by means of a simple equation in the form $Y = a + bX$. In this equation Y is the value of the dependent variable on the Y axis; X is the value of the independent variable on the X axis. a is the value of the intercept on the Y axis when X is zero; b is the regression coefficient. The regression coefficient is the number of units that the value of Y changes for every unit change in the value of X. As can be seen in Figure 1.33, the regression coefficient describes the slope of the graph line. In Figure 1.33(a) the equations are all very simple because when X is zero Y is also zero. Hence we only need to be concerned with the slope of the graph line, which is indicated by the regression coefficient. The 2 in $Y = 2X$ means that Y changes by two units for every unit change in X. $Y = X$ means that Y changes by one unit for every unit change in X. The regression coefficient is 1.0. The 0.5 in $Y = 0.5X$ means that Y changes by half a unit for every unit change in X. Clearly it is the regression coefficient that governs the slope of the graph line. In Figure 1.33(b) there is an intercept of two units on the Y axis. Hence all the equations have a value of 2 for 'a'. As in Figure 1.33(a) the regression coefficient indicates the slope of the graph line. In Figure 1.33(c) the intercept on the Y axis when X is zero is negative. This makes no difference at all to the rules about the equation. The value for 'a' in the general equation simply becomes negative as in $Y = -2 + X$ and $-3 + 0.5X$. The regression coefficient still governs the slope of the line. All the graph lines and equations so far have represented a positive relationship. As the value of X has increased, so has the value of Y. In Figure 1.33(d) however the relationship is negative. As the value of X increases the value of Y decreases. On the regression line for $Y = 10 - X$, when Y is 9 X is 1 and when Y is 1 X is 9. The regresssion coefficient is still calculated in the same way as before but it is preceded by a minus sign instead of a plus sign.

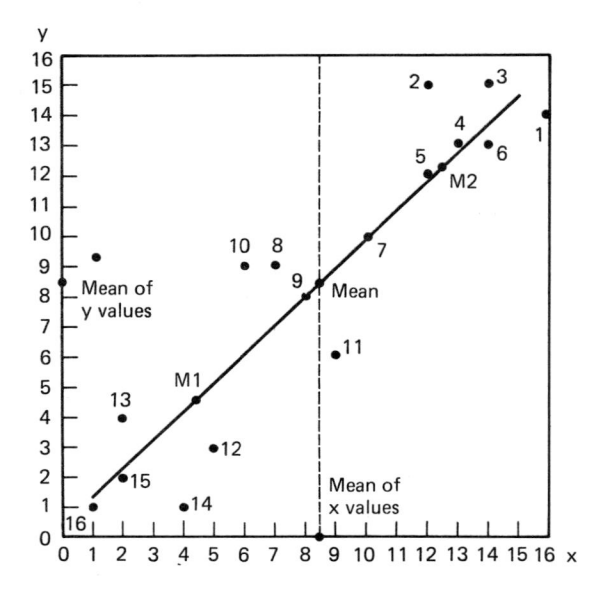

Fig. 1.32 The construction of a regression line

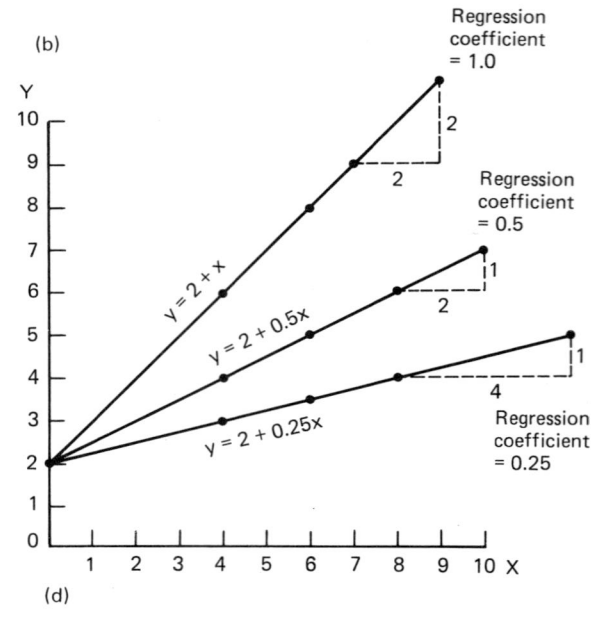

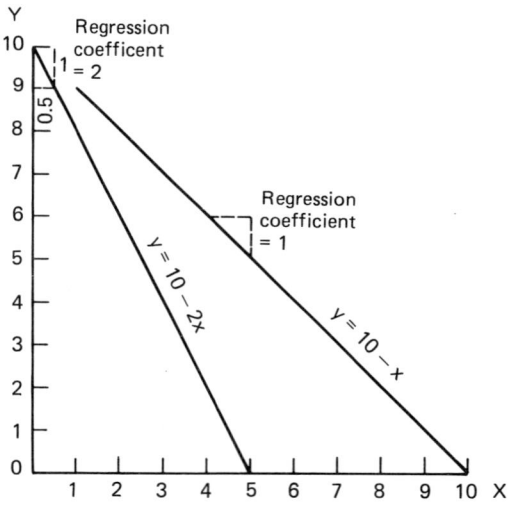

Fig. 1.33 Equations of a straight line

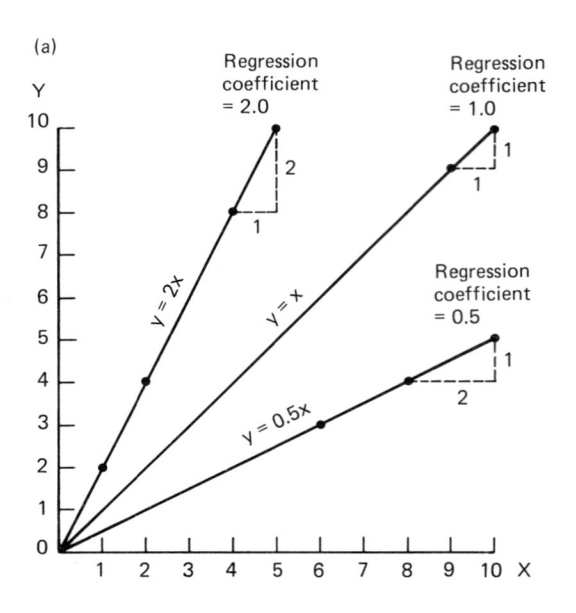

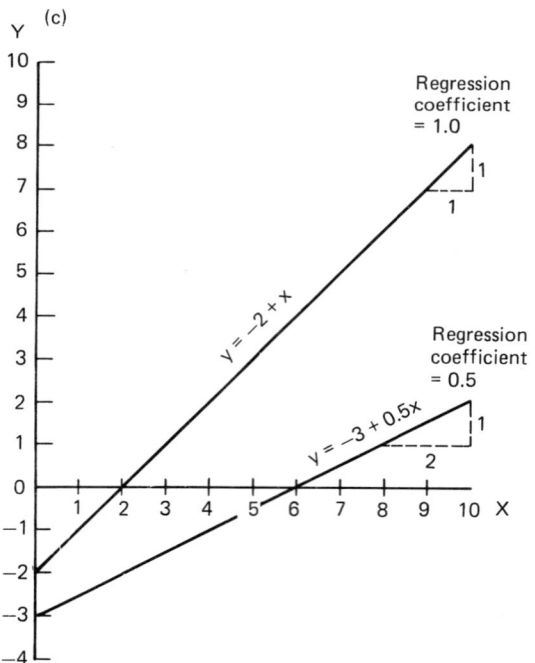

Exercises

1. Look at the data in the box opposite.
    (a) Construct a 'simple array' and an 'array with frequencies'.
    (b) Construct a 'frequency graph' in the same style as Figure 1.1 (*e*) and (*g*) for each of the following groupings of frequencies:
       (i)   13 – 15; 16 – 18; 19 – 21; 22 – 24; 25 – 27; 28 – 30.
       (ii) 13 – 17; 18 – 22; 23 – 27; 28 – 32.
       (iii) 13 – 14; 15 – 16; 17 – 18; 19 – 20; 21 – 22; 23 – 24; 25 – 26; 27 – 28.
    (c) Compare the efficiency of the three graphs in illustrating the characteristics of this frequency distribution.

GRAPHS

2. (a) Using the data given below calculate the cumulative frequency for the successive classes.

Class range	Class frequency	Cumulative frequency
Less than 9.5		0
9.5–under 12.5	2	
12.5–under 15.5	5	
15.5–under 18.5	8	
18.5–under 21.5	8	
21.5–under 24.5	5	
24.5–under 27.5	2	

   (b) On graph paper, using the cumulative frequency distribution, construct an ogive.

   (c) Comment on the shape of the ogive you have drawn and explain why it differs from the ogives shown in Figure 1.4.

3. Tables A and B below have been compiled by measuring the length of the intercepts between contour lines along a series of different transects running across a contour map. Table A shows the length of the intercepts in the various altitude zones. It is assumed that the length of the intercepts is proportional to the area of the altitude zone. Table B lists the percentage of the total area that is occupied by each altitude zone according to the above assumption. Illustrate the characteristics of the relief of this area by means of
   (a) a histogram;
   (b) an ogive.

4. Value of Agricultural Production in New England in 1982 (millions of dollars)

	Livestock	Crops
Connecticut	183.5	102.8
Maine	256.7	143.0
Massachusetts	142.3	139.5
New Hampshire	76.8	26.2
Rhode Island	12.3	18.1
Vermont	349.5	20.1
Total	1021.1	449.7

   (a) Construct two pie graphs, similar to those in Figure 1.9, showing the differences between the distributions of livestock production and crop production in New England.

   (b) Referring to your pie graphs write a short comparison between the distributions of livestock production and crop production in New England.

5. The table on page 34 describes the climate of a location in the Far East. Draw a pair of circular graphs (one for temperature and the other for rainfall) to illustrate the characteristics of this climate.

Data for question 1. Mean temperatures (°C) of 30 locations in the Tropics for the month of January.

27	25
28	27
27	28
23	23
24	18
18	24
13	19
20	24
22	21
21	25
22	21
21	16
26	16
18	24
24	25

Table A Length of intercepts on contour map (mm)

Altitude zone	A	B	C	D	E	F	G	H	I	J	Total
300–350 m	0	0	0	0	0	13	0	0	0	0	13
250–330 m	0	0	0	16	51	73	31	22	19	22	234
200–250 m	23	14	54	96	83	78	81	18	26	26	499
150–200 m	60	127	117	74	56	26	78	61	104	48	751
100–150 m	94	49	19	4	0	0	0	62	41	78	347
50–100 m	13	0	0	0	0	0	0	27	0	16	56
Total	190	190	190	190	190	190	190	190	190	190	1900

Table B Percentages of area in altitude zones

Altitude zone	Total length of intercepts	% of total area	Cumulative % of area
300–350	13	0.68	0.68
250–300	234	12.32	13.0
200–250	499	26.26	39.26
150–200	751	39.53	78.79
100–150	347	18.26	97.05
50–100	56	2.95	100.00

	Mean temperature (°C)	Mean rainfall (mm)
January	3.0	56
February	4.0	71
March	7.0	112
April	13.0	125
May	17.0	145
June	21.0	165
July	24.0	135
August	25.0	145
September	22.0	221
October	16.0	188
November	10.0	107
December	5.0	53

(a) Construct a pair of circular graphs (like those in Figures 1.10 and 1.11) to illustrate the characteristics of this climate.

(b) Write a comparison of the climate shown in the circular graphs you have drawn with the Mediterranean climate shown in Figures 1.10 and 1.11.

(c) Plot the mean monthly temperatures in the table above as a simple line graph. Compare the advantages and disadvantages of the circular graph and the simple line graph for describing temperature regimes.

6. The table below lists mean monthly temperatures and precipitation totals for a location in England and a location in Australia.

	Mean monthly temperature (°C)		Mean monthly precipitation (mm)	
	England	Australia	England	Australia
January	5	22	48	94
February	5	22	38	102
March	6	21	43	122
April	8	18	38	142
May	11	15	51	130
June	13	12	53	122
July	15	11	58	122
August	15	13	76	76
September	14	15	43	74
October	11	17	74	81
November	8	19	56	71
December	6	21	58	74

(a) Use this table and a sheet of graph hythergraphs in a similar style to Figure 1 stations.

(b) Compare the climates of these two locations.

7. The table below lists the maximum daily temperatures (°C) over a period of 30 days.

15 12 14 10 11 15 18 20 22 19
14 16 18 18 20 21 17 22 24 20

(a) Plot each of these temperatures on a sheet of graph paper and draw a line joining all the individual values to show the daily fluctuations.

(b) On the same graph plot the 3-day running means and the 5-day running means.

(c) Comment on the similarities and differences between the three line graphs you have drawn.

(d) Write a description of the temperature changes that occurred over the 30 days.

8. A study of two areas yields the following information about the relationships between area and altitude.

Altitude range (m)	Area of land within altitude range stated (km^2)	
	Area A	Area B
400–500	100	100
300–400	250	600
200–300	300	500
100–200	200	200
0–100	150	600
	1000	2000

(a) On a sheet of graph paper plot a hypsometric curve for each of these areas. Refer to Figure 1.18(c).

(b) Compare the characteristics of the relief of these two areas.

9. The tables below describe aspects of farming in New England.

Percentage of New England's total area of farmland

Connecticut	10
Maine	32
Massachusetts	13
New Hampshire	10
Rhode Island	1
Vermont	34
	100

GRAPHS

Percentage of New England's dairy production

Connecticut	14
Maine	14
Massachusetts	12
New Hampshire	8
Rhode Island	1
Vermont	51
	100

Value of New England's crop production (millions of dollars)

Connecticut	102.8
Maine	143.0
Massachusetts	139.5
New Hampshire	26.2
Rhode Island	18.1
Vermont	20.1
	449.7

(a) Construct a Lorenz curve to show the degree of specialisation in dairying in relation to the total area of farmland.

(b) Construct a Lorenz curve to illustrate specialisation in crop production in relation to total area of farmland. In this case it is necessary to calculate percentage shares of the various states in New England's crop production.

10. (a) Study Figure 1.23(b) which shows the example of the rank-size rule. Estimate from the graph the population of:
   (i)   The sixth ranked city;
   (ii) The twelfth ranked city;
   (iii) The fiftieth ranked city;
   (iv) The two hundredth ranked city.
   Check your answers by arithmetic.

(b) Using a sheet of semi-logarithmic graph paper, draw a graph to illustrate the characteristics of a drainage pattern which has one 5th order stream and the number of lower order streams increases by a factor of 4 for each successive lower order.

(c) Use semi-logarithmic graph paper to determine which of the following sequences represents a constant rate of population change.

Population total (thousands)

1981	1982	1983	1984	1985	1986
5	10	15	20	25	30
10	30	50	70	90	110
1	2	3	4	5	6
4	8	16	32	64	128

11. The table below shows the result of a map study of land utilisation in five different rural areas, A, B, C, D and E.

	Percentage of the area:		
	Arable	Grassland	Rough grazing
Area A	33	10	57
Area B	24	38	38
Area C	10	22	68
Area D	28	42	30
Area E	48	20	32

Plot these values on a triangular graph and, referring to Figure 1.30, describe the similarities and the differences in land utilisation between these five areas.

12. (a) Draw graphs to represent each of the following equations:

$Y = 3X$
$Y = 6 + 2X$
$Y = 20 - 3X$
$Y = 10 + 0.5X$
$Y = 4 + X$
$Y = -4 + X$

(b) State the equation for each of the regression lines, A to E, shown below.

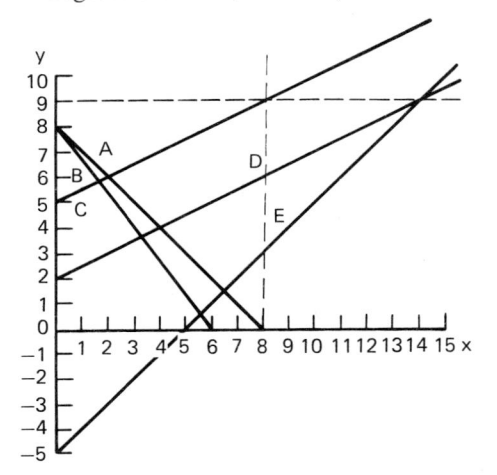

2 Maps

Graphs are particularly well suited to illustrating variations in time and space of geographical characteristics such as altitude, climatic regimes, birth rates, crop production and employment categories. They can illustrate changes that occur both from area to area and from time to time. The quantitative relationships between two or even three variables can be depicted with great clarity. Graphs, however are usually concerned with quantities, percentages and frequencies, and not with the spatial distributions that are so important in geography. Maps, on the other hand, are primarily concerned with the spatial distribution of geographical phenomena. Maps can be regarded as the fundamental tools of geographers and also a most important product of their work.

2.1 Dot maps

A dot map is a map in which the spatial distribution of a geographical variable is represented by a number of dots of equal size. Figure 2.1(a) is such a map showing the distribution of population in southern England and South Wales.

The construction of a dot map
Figure 2.1(a) was constructed in the following way. First of all the outlines of the county boundaries were drawn. Then from the county population totals it was decided that if each dot represented a population of 100 000 there would be a satisfactory density of dots and excessive sparseness and overcrowding would be avoided. Also it was necessary to take account of the proposed size of each dot. The number of dots to be allocated to each county area was determined by dividing the county's total population (in thousands) by 100. This value invariably needed to be rounded. For example, a county population of 87030 was allocated 9 dots. This degree of inaccuracy seems inevitable when working at this scale.

The next step was to distribute the dots as accurately as possible within each county. To do this the population totals of the various smaller administrative areas within the county needed to be considered. Urbanised administrative areas with more than 100 000 people could each be allocated at least one dot. Bristol, for example, was allocated a cluster of four dots and Southampton was allocated two.

The remaining dots could not be placed quite so accurately. Each needed to be located as far as possible centrally within an area where a generally rural population reached a total of 100 000. This could be quite a large area. More accurate details of this kind would have resulted if a dot had been made to represent fewer people, but this would have increased the total number of dots so that congestion could only have been avoided by increasing the size of the map or reducing the size of the dots.

An advantage associated with a dot map is that it is fairly easy to find the approximate total population of the various counties simply by counting the dots. Differences in population density can to some extent be deduced by relating the number of dots to the size of each county. There may however be a tendency to regard blank areas as completely unpopulated.

Figure 2.1(b) is identical to Figure 2.1(a) except that the county boundaries have been omitted. This seems to make it easier to form an impression of the broader variations in population density over the whole area. Relatively densely populated areas are clearly recognised on both sides of the Bristol Channel, along the south coast east of Southampton and along the northern margins of Surrey and Berkshire.

MAPS 37

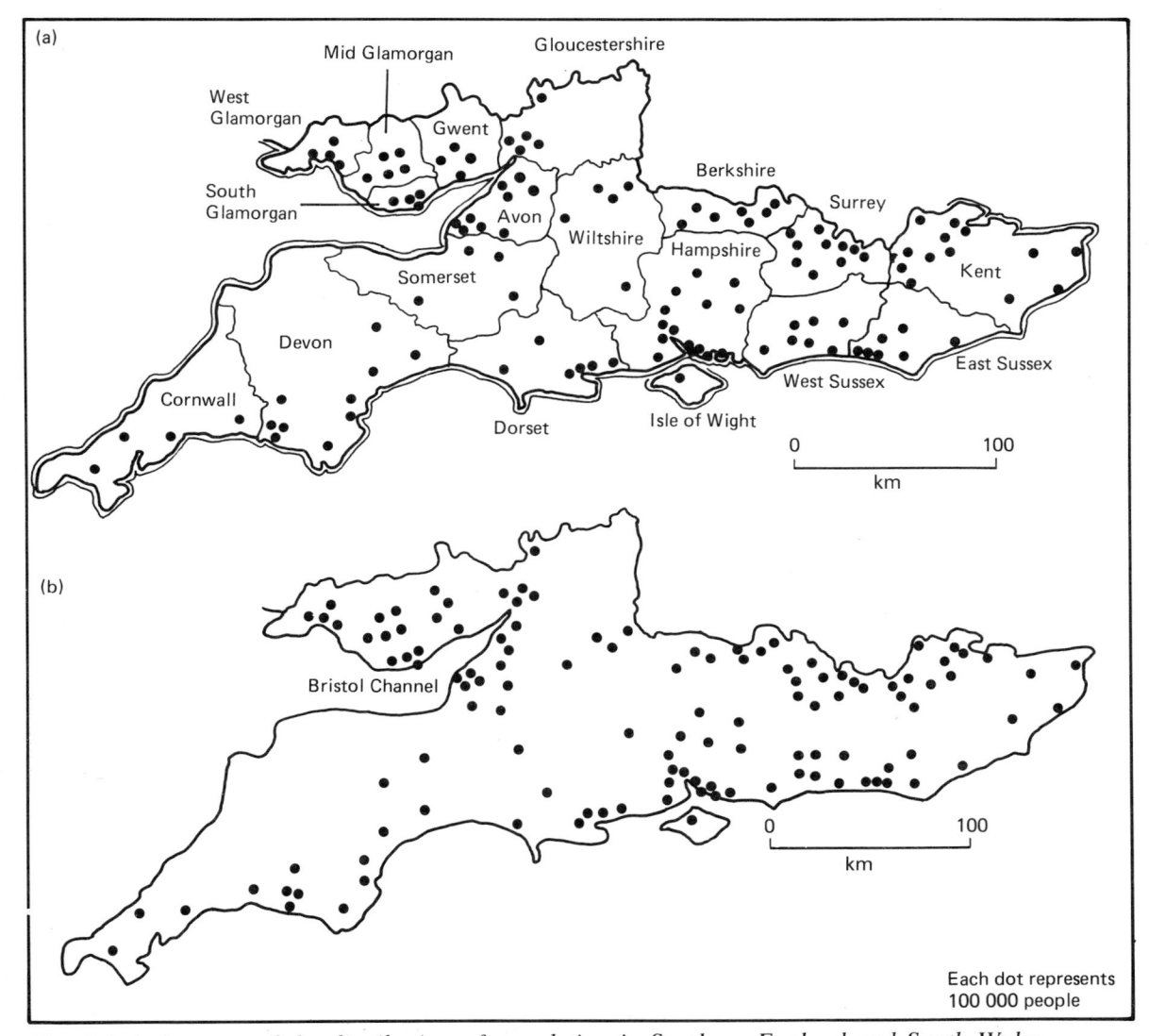

Fig. 2.1 A dot map of the distribution of population in Southern England and South Wales

2.2 Choropleth maps

A choropleth map is one in which spatial distributions are represented by showing overall differences and similarities between administrative areas. For each administrative area information is provided in the form of some kind of average value, such as density of population, percentage of land under a particular crop, average crop yield per unit area, percentage of the population employed in particular industries. It is not possible to read from a choropleth map information about total values, as can be done from a dot map. In some cases it may be possible to deduce information about total values very approximately. A very large administrative area with a high population density, for example, must have a greater population than a small administrative area with a low population density, but it would be difficult to deduce actual population totals. There would be little point in representing total population on a choropleth

County	Population density (persons per ha)	Large differences between successive densities
Cornwall	1.2	
Somerset	1.3	
Devon	1.4	
Wiltshire	1.5	0.4
Gloucestershire	1.9	0.4
Dorset	2.3	0.8
Isle of Wight	3.1	
Gwent	3.2	
West Sussex	3.4	
East Sussex	3.7	
Hampshire	3.9	
Kent	4.0	0.5
West Glamorgan	4.5	0.8
Mid Glamorgan	5.3	
Berkshire	5.6	0.4
Surrey	6.0	0.9
Avon	6.9	2.5
South Glamorgan	9.4	

Density Classification A

0 – 1.9	Cornwall, Somerset, Devon, Wiltshire, Gloucestershire
2 – 3.9	Dorset, Isle of Wight, Gwent, West Sussex, East Sussex, Hampshire
4 – 5.9	Kent, West Glamorgan, Mid Glamorgan, Berkshire
6 – 7.9	Surrey, Avon
8 – 9.9	South Glamorgan

Density Classification B

1 – 2.9	Cornwall, Somerset, Devon, Wilts, Gloucestershire, Dorset
3 – 4.9	Isle of Wight, Gwent, West Sussex, East Sussex, Hampshire, Kent, West Glamorgan
5 – 6.9	Mid Glamorgan, Berkshire, Surrey, Avon
7 – 8.9	—
9 – 10.9	South Glamorgan

Density Classification C

1 – 1.9	Cornwall, Somerset, Devon, Wiltshire, Gloucestershire
2 – 2.9	Dorset
3 – 3.9	Isle of Wight, Gwent, West Sussex, East Sussex, Hampshire
4 – 4.9	Kent, West Glamorgan
5 – 5.9	Mid Glamorgan, Berkshire
6 – 6.9	Surrey, Avon
7 – 7.9	—
8 – 8.9	—
9 – 9.9	South Glamorgan

Density Classification D

1 – 2.3	Cornwall, Somerset, Devon, Wiltshire, Gloucestershire, Dorset
2.4 – 4.5	Isle of Wight, Gwent, West Sussex, East Sussex, Hampshire, Kent, West Glamorgan
4.6 – 6.0	Mid Glamorgan, Berkshire, Surrey
6.1 – 6.9	Avon
7.0 and over	South Glamorgan

Fig. 2.2 The classification of the counties of South Wales and Southern England by population density

MAPS

map because all the large administrative areas would have greater totals than the small areas even with a constant density. A large area with a low population density could have a greater total population than a small area with a high population density.

The construction of a choropleth map

In constructing a choropleth map of the distribution of population in an area it is necessary first of all to calculate the average density for each administrative area by dividing its total population by its area in hectares. This gives the density of population per hectare. Next the administrative areas are grouped into classes, usually between about five and eight, according to the population density. These classes are then drawn on the map. Boundaries between administrative areas belonging to the same class may be omitted (Fig. 2.3).

Even though the population distribution in each area is not uniform it is permissible to

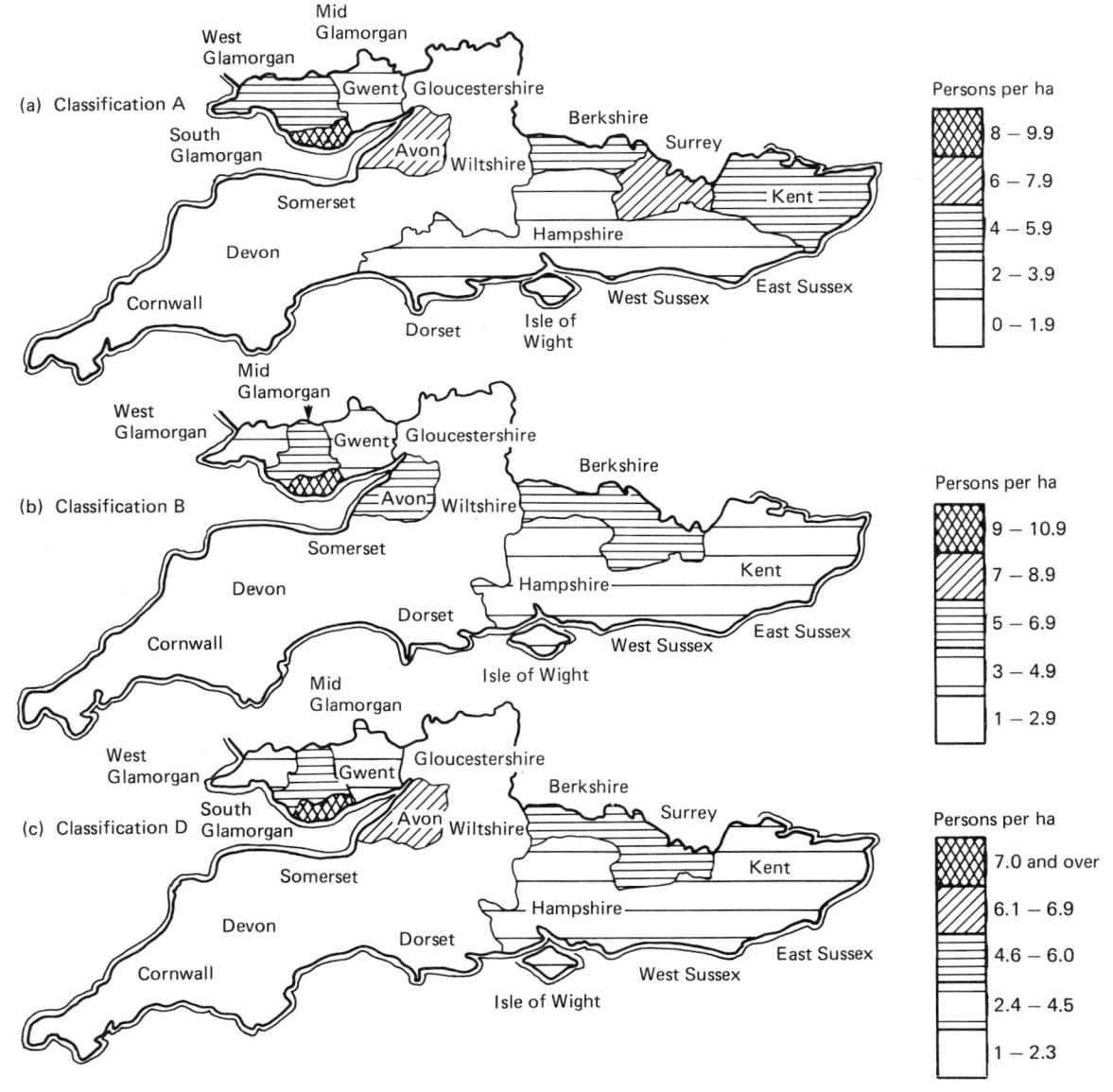

Fig. 2.3 Choropleth maps

shade the whole of each area in a similar style. This is so because the shading indicates the density for the corresponding area. No information is provided about variations within a single administrative area. A weakness of choropleth maps in comparison with dot maps is that they give the impression that the density changes suddenly at the perimeter of each area.

Figures 2.2 and 2.3 illustrate the construction of a set of choropleth maps, using the same data as was used for the dot maps in Figure 2.1. First of all, in Figure 2.2, the counties have been listed in rank order of population density from Cornwall (the lowest) to South Glamorgan (the highest). Cornwall has a population density of 1.2 per ha. South Glamorgan has a population density of 9.4 per ha. The range of population densities is therefore 9.4 − 1.2 = 8.2. It appears therefore that a class interval of about 2 per ha should yield about five different classes.

Classification A demonstrates this. It provides 5 counties in the 0 − 1.9 class, 6 in the 2 − 3.9 class, 4 in the 4 − 5.9 class, 2 in the 6 − 7.9 class and one in the 8 − 9.9 class. Classification B also has a class interval of 2 per ha but it begins at a higher density. This classification has 6 in the 1 − 2.9 class, 7 in the 3 − 4.9 class, 4 in the 5 − 6.9 class, none in the 7 − 8.9 class and one in the 9 − 10.9 class. Because of the wide gap between population densities of Avon (6.9) and South Glamorgan (9.4) there is an empty class. Classification C has 9 classes each with an interval of one person per ha but two of these classes are empty because of the large gap between the densities of Avon (6.9) and South Glamorgan (9.4). Of these classes with equal class intervals therefore Classification A seems to be the most satisfactory.

The class intervals in Classification D have been determined differently. An attempt has been made to match the class boundaries with large differences between successive population densities. The class intervals therefore vary in size. The first class boundary is placed between Dorset and Isle of Wight where there is a difference in density of 0.8 (Fig. 2.2). The next is placed between West Glamorgan and Mid Glamorgan (a difference in density of 0.8). The next comes between Surrey and Avon (difference 0.9) and the last between Avon and South Glamorgan (difference 2.5). Classifications A and D therefore appear to be the

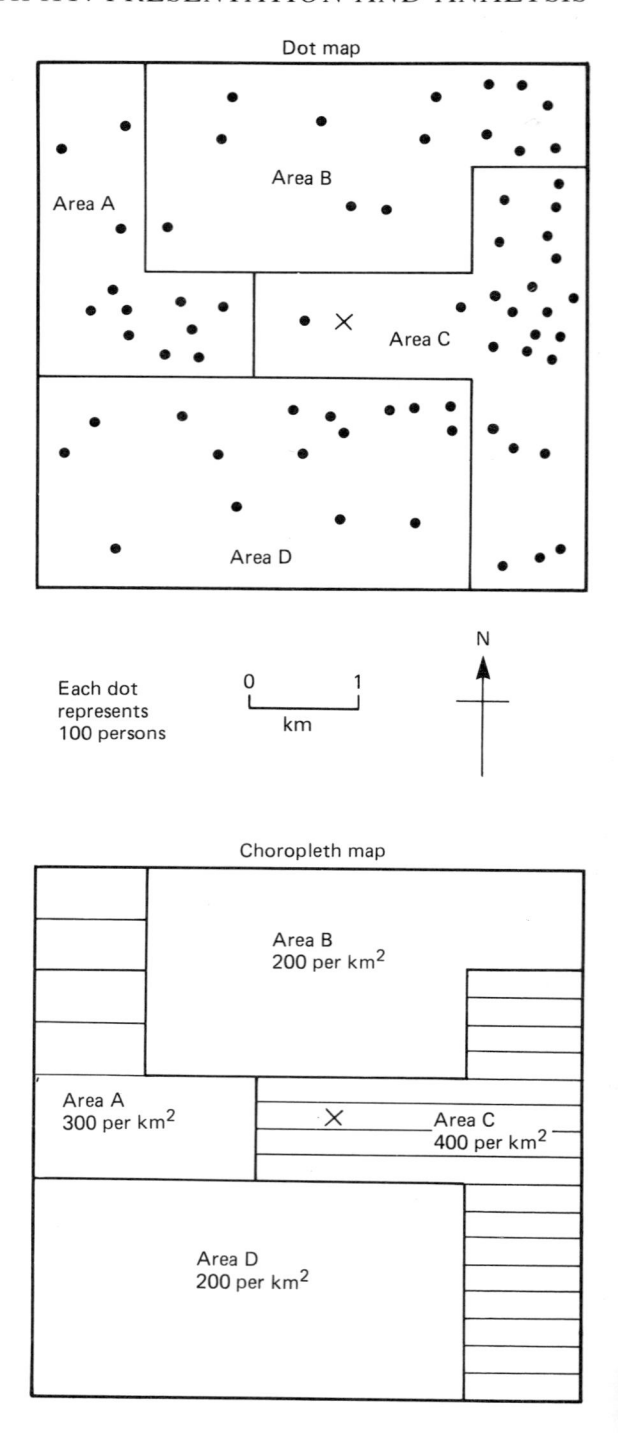

Fig. 2.4 Dot and choropleth maps

most useful. The table below shows their similarity.

Classification A		Classification D	
	Number of cases		Number of cases
0–1.9	5	1.0–2.3	6
2–3.9	6	2.4–4.5	7
4–5.9	4	4.6–6.0	3
6–7.9	2	6.1–6.9	1
8–9.9	1	7.0 and over	1
	18		18

On the whole the choropleth maps based on Classifications A and D (Fig. 2.3(a) and (c)) correspond well with the dot maps (Fig. 2.1(a) and (b)). The sparsely populated belt extending from Cornwall, through Devon, Somerset and Wiltshire to Gloucestershire is clearly seen in Figure 2.3(a) and Dorset is also included in Figure 2.3(c). The relatively densely populated South Glamorgan and Avon are also clearly shown. From Hampshire eastwards population densities are generally moderate but they tend to increase in the north in Berkshire and Surrey. However, only Classification A (Fig. 2.3(a)) identifies the greater density in Surrey than in Berkshire.

Sometimes it is possible for a choropleth map to give a false impression of the distribution of population densities. Figure 2.4 shows a dot map and a choropleth map of the distribution of population in four imaginary areas, A, B, C and D. On the dot map each dot represents 100 persons. There are 12 dots in Area A, which occupies 4 km². Each dot represents 100 persons. So the population density of Area A is $\frac{12 \times 100}{4}$ which is 300 per km². The population densities of the other areas have been calculated in the same way in order to produce the choropleth map. When the two maps are compared some 'anomalies' appear. On the choropleth map Area C has a population density of 400 per km², the highest density of all the areas. Care must be taken however not to interpret this too literally. It does not mean that *all parts* of Area C have the highest density. At the point marked X on the choropleth map there appears to be a population density of 400 per km², which is considerably greater than that of Areas A, B and D. However, when we look at the dot map it is quite clear that there is a much higher population density to the south, in Area D, which on the choropleth map has a density of only 200 per km². In fact the area near X contributes very little towards the high population density of Area C. It just happens to have been included in the administrative area known as 'Area C'. Most of Area C's people live in the east of the area and there are enough people there to give Area C the highest *average* density even though it contains some relatively sparsely populated districts.

2.3 Proportional symbols

Proportional symbols indicate numerical values on maps at particular points or in particular areas. Proportional circles are rather similar to the dots used in a dot map but the circles vary in size according to the differing quantities represented. Figure 2.5 shows by the use of proportional symbols the information that is shown on the dot maps and choropleth maps of Figures 2.1 and 2.3. The area of each circle is proportional to the population of the corresponding county.

Proportional symbols can be used in both choropleth maps and dot maps. In choropleth maps they can show a quantity that refers to an area such as a county or a country. This can be total population, as in Figure 2.5, or total weight of crops or minerals produced, or any other total value for an administrative area. In addition to showing total values a proportional symbol can be subdivided to show various subdivisions of the total. In this case each proportional symbol is a pie graph if the proportional symbol is a circle.

Proportional symbols can also be used in dot maps. Instead of allocating a constant value to each dot on the map as in Figure 2.2, proportional symbols varying in value can be inserted in the correct locations. These symbols can show, for example, the total populations of each city, the total trade of each port or the number of flights associated with each airport.

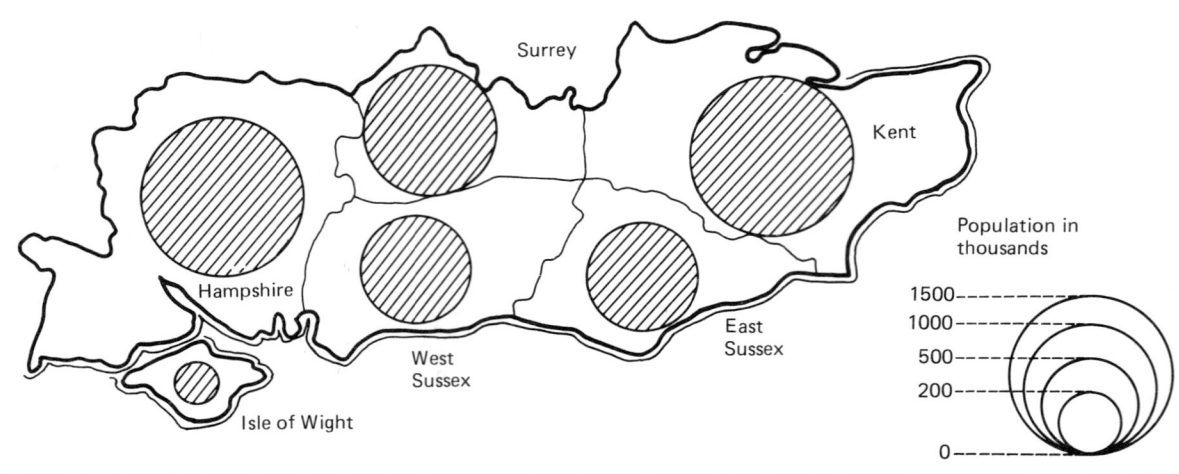

42 PRACTICAL GEOGRAPHY: PRESENTATION AND ANALYSIS

Fig. 2.5 Population distribution in the counties of south-east England shown by proportional circles

(a) Proportional bars

(b) Proportional circles

(c) Proportional squares

(d) Proportional cubes

(e) Proportional spheres

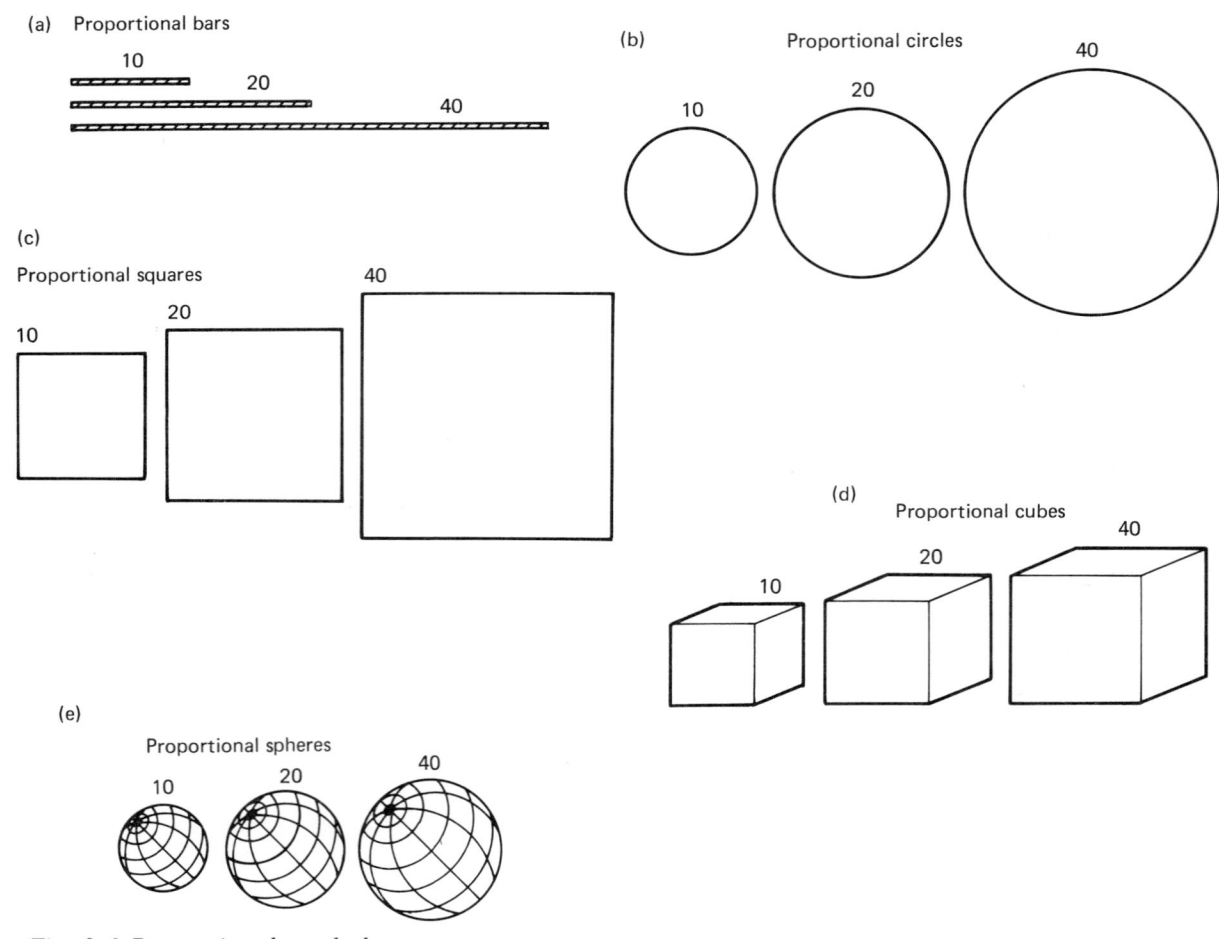

Fig. 2.6 Proportional symbols

Types of proportional symbol

Figure 2.6 shows some of the most common types of proportional symbol. A very simple type is the *proportional bar* (Fig.2.6(a)). This consists of a bar whose *length* is proportional to the quantity represented. The bar may be aligned either vertically, horizontally or obliquely. Its chief advantages are that it is easy to draw and that its simple linear scale is easy to understand and use. Its main drawback is that the rapid increase in its length makes it difficult to fit into administrative areas that have a compact shape. Its length increases much faster than that of any of the other symbols shown in Figure 2.6. It could be useful for showing port statistics, in which case the bar might be able to extend into the sea.

In *proportional circles* it is the *area* of the symbol that is made proportional to the quantity represented. The area of a circle is πr^2. But in this case π need not be taken into account because it is a constant. Hence it is only necessary to find the square root of the quantities to be represented, and these square roots are used as the radii of the proportional circles. This will ensure that the areas of the circles are proportional to the quantities represented.

In Figure 2.6(b) the quantity to be represented by the first circle was 10. The square root of 10 is 3.16. The absolute size of the circle depends on the scale that is used to draw the radius of 3.16 units. The circle could be drawn with a radius of 3.16 mm (taking 1 mm as a unit), or with a radius of 3.16 cm (taking 1 cm as a unit) or with a radius of 1.58 cm (taking 0.5 cm as a unit). Provided that all the other circles, whatever their size, are constructed from radii calculated using the same unit they will all be in correct proportion.

The use of proportional circles seems to be a very efficient way of showing distributions. Circles are quite easy to draw and the circle is a very compact shape which tends to fit snugly into administrative areas (see Figure 2.5). The circle has the maximum possible area and the minimum possible perimeter. It is also possible to overlap proportional circles in areas where they are clustered. This can provide a quite desirable appearance of congestion.

Proportional squares are somewhat similar to proportional circles, but they are less compact and they cannot be divided internally so satisfactorily. The area of the symbol is made proportional to the quantity represented by making each side proportional to the square root of the total quantity to be represented.

Proportional cubes (Fig. 2.6(d)) use all three dimensions. The volume of the cube is made proportional to the quantity represented by making each side of the square front face proportional to the cube root of the quantity to be represented. In Figure 2.6(d) a quantity of 10 units is represented by a cube whose front face is bounded by sides measuring 2.15 (the cube root of 10) each. In the cube which represents 20 units each side of the front face measures 2.71 units (the cube root of 20). The faces on the top and at the side of the cube are not drawn to scale. They are drawn so as to give the impression of a regular cube.

Proportional spheres (Fig. 2.6(e)) are rather similar to proportional cubes. The volume of the sphere is made proportional to the quantity represented by making the radius of the sphere proportional to the cube root of the quantity represented. A three-dimensional effect is achieved on each symbol either by drawing a pattern of meridians and parallels as in Figure 2.6(e) or by shading them black and inserting a 'highlight' near the top of the sphere.

Because they are 'solid' figures and represent a 'volume', proportional spheres and cubes can depict a very wide range of values with relatively little change in the size of the symbol (Figure 2.6). Quite a small increase in the length of a cube's side or a sphere's radius results in a considerable increase in volume. They therefore make it easier to depict large quantities and to fit a symbol into a limited space on a map. Although this is an advantage in one sense it is also a disadvantage in that it becomes more difficult for the reader to make an estimate of the quantity that is represented by the symbol. In Figure 2.6 all the sets of symbols represent a sequence with successive quantities of 10 units, 20 units and 40 units. In each case the quantity represented doubles from left to right. The proportional bars are clearly the easiest to use for estimating quantities but they are the most difficult to represent on a map. The proportional spheres are the most compact and use relatively little map space, but they do not give a clear indication of absolute quantities.

2.4 Isoline maps

An isoline (isopleth, isarithmic) map is one in which lines (isolines) have been drawn to show the variations in the values represented. A single isoline connects the points on the map at which the variable represented has an equal value. Contour maps, for example, have isolines (contour lines) connecting points of equal altitude. On maps depicting weather or climate isotherms connect points of equal temperature; isobars connect points of equal pressure; and isohyets connect points of equal precipitation. These variables, altitude, atmospheric pressure and precipitation, are continuously distributed over the earth's surface. It is therefore appropriate to represent them as an isoline map which is based on the assumption that the distribution represented is that of a spatially continuous variable. Dot maps (page 36), choropleth maps (page 39) and proportional symbols (page 41) are also sometimes used to depict what are virtually spatially continuous variables such as population (Figs. 2.2 and 2.3) but these techniques tend to divide geographical space into point locations in the case of dot maps and separate areas in the case of choropleth maps.

The construction of an isoline map
In constructing an isoline map the first step is to plot the locations of as many values of the variable as possible. Figure 2.7(*a*) is a theoretical example of the type of diagram that is required. More often, however, the locations will be distributed irregularly, as in the case of spot heights or weather recording stations.

Next the size of the interval between successive isolines is decided. In Figure 2.7 the selected interval is 5 units. Isolines are to be drawn for successive values of 5, 10, 15, 20 and 25. It is best to begin by plotting the isoline for a value of either 5 (the lowest) or 25 (the highest). If we start with '25' isoline our attention is drawn to the south-west corner of the map. Here we can see the only point with an exact value of 25. The isoline must pass through this point. To the west of this point is a value of 28 and to the north a value of 22. 25 is midway between these, so the isoline should pass half-way between them. To the east of the 25 point we see values of 23 and 28. The '25' isoline should pass between these, slightly nearer to the 23 point than to the 28 because 23 differs from 25 by only 2 units and 28 differs by 3 units.

It is not possible to produce an *exactly* accurate map. We have no detailed information

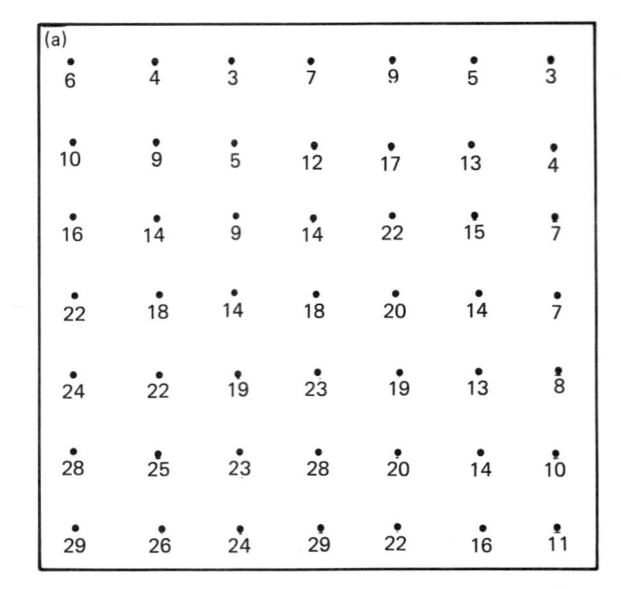

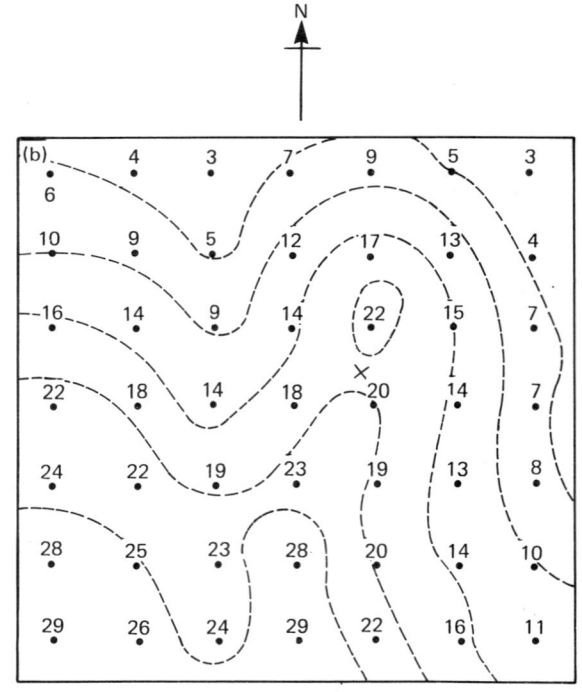

Fig. 2.7 The construction of an isoline map

MAPS

about the values that lie between any pair of neighbouring points. We do know that between the 25 point and the 28 point there must be examples of all the values between 25 and 28 but we cannot tell what variations actually occur.

There may be a sudden increase to almost 28 near the 25 point or the value may rise suddenly near the 28 point. There is also a problem near the point marked X (Fig.2.7(b)) near the centre of the map. Here it has been assumed that point

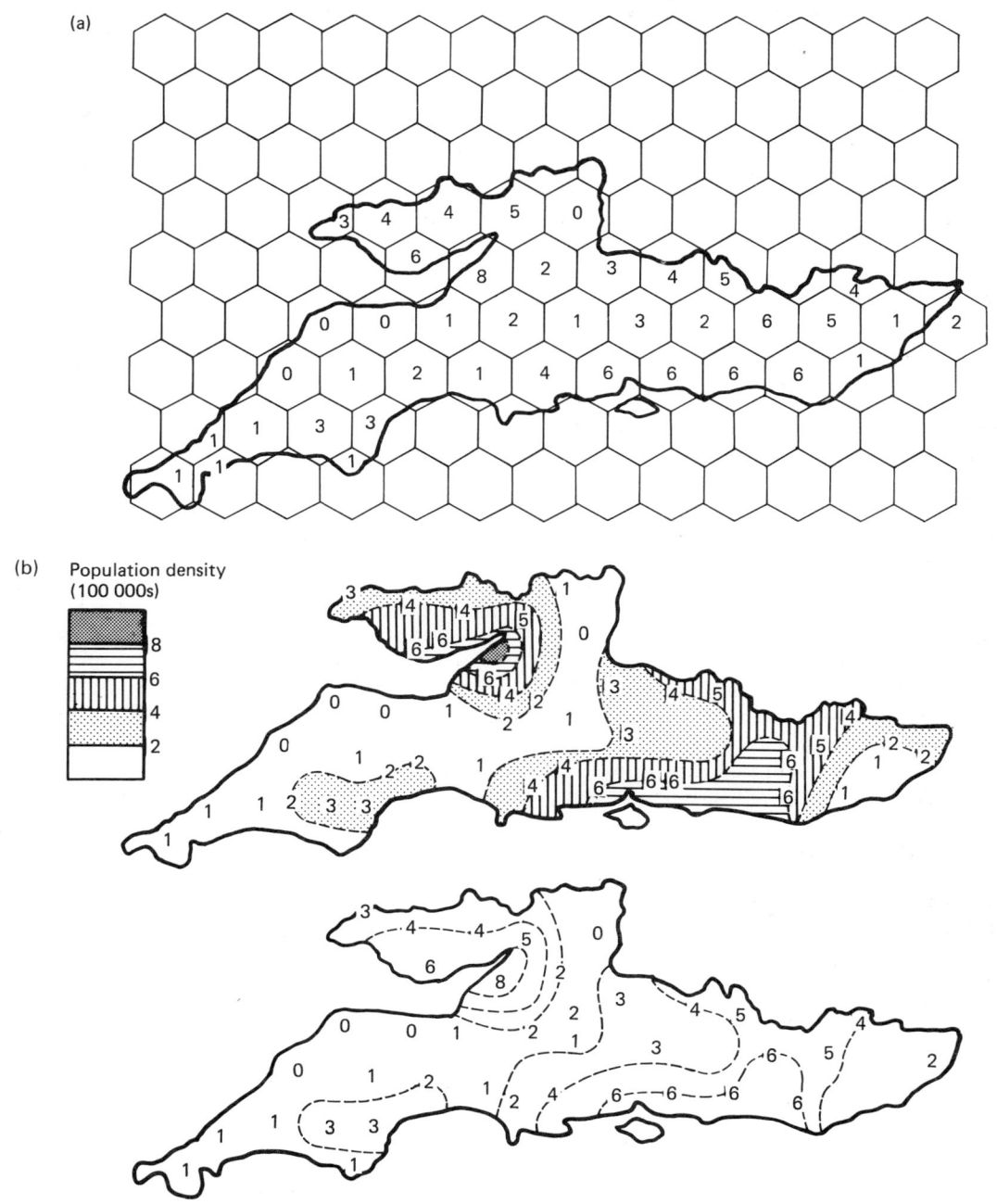

Fig. 2.8 The construction of an isoline map from a dot map

X has a value of less than 20, so a closed oval has been drawn round the 22 point. If X has a value of over 20 the closed oval merges with the 'over 20' area to the south. One has to accept therefore that isoline maps cannot be completely accurate, but the errors should not be serious because the variables they are used to depict are spatially continuous.

Figure 2.8 illustrates a method of transforming a dot map (Fig. 2.1(b)) into an isoline map. First of all the dot map is covered by a set of hexagons of equal size. Next, the number of dots occurring within each hexagon is recorded at the centre of the hexagon (Fig. 2.8(a)). Each dot represents 100 000 people (Fig.2.1(a)). Since the hexagons are all of equal size the number recorded at the centre of each hexagon is a measure of its density of population. In Figure 2.8(b) the outlines of the hexagons have been omitted and the numbers have been treated as 'spot heights'. Isolines representing 2, 4, 6 and 8 (100 000s) have been fitted to the hexagonal pattern of population density values. The isoline map so produced gives a pleasing impression of the spatial variations in population density and is probably easier to understand than the dot map from which it is derived. The relatively high densities in Avon near the Bristol Channel and along the coast of Hampshire and West Sussex show up clearly. Berkshire, Surrey and Glamorgan-Gwent show as moderate densities. The sparsely populated belt extending through Cornwall, Devon, Somerset, Wiltshire and Gloucestershire, is also clearly shown, as is the relatively low density in south-eastern Kent. A criticism might be that the cluster of dots in Berkshire and Surrey seems slightly undervalued. This could be because the hexagons (Fig. 2.8(a)) in these areas extend beyond the study area so that dots can only exist in part of the hexagon.

It is also possible to construct an isoline map of population densities by using the population density (persons per ha) data in Figure 2.2. In this case, instead of hexagons, the actual pattern of counties is used. The population density value is plotted at one point within each county area, either in the centre of the area or nearer to the part of the county that has the highest population density. This produces a much coarser map than Figure 2.8(b) since each county is fairly large in comparison with the whole study area. This method however, could be useful when dealing with a large number of relatively small administrative areas.

Isotims

Isolines are also used in theoretical locational analysis in the form of isotims (isovectures) which join points of equal transport cost for either a raw material or a finished product, and isodapanes which join together points with equal *total* transport costs. Figure 2.9 illustrates a simple example of the use of isotims and isodapanes. Isolines of transport costs are used to determine the point at which transport costs are least and also to describe the pattern of transport costs over an area.

It is assumed that an industry uses 2 tonnes of a single raw material to produce 1 tonne of product. The source of the raw material is in a different location from the market to which the product is delivered.

Figure 2.9(a) shows the relative locations of the raw material and the market which are 3 km apart. Since 2 tonnes of raw material are needed to manufacture 1 tonne of product, it is necessary to transport 2 tonnes of raw material to the market to be manufactured into 1 tonne of product. The distance between the raw material and the market is 3 km, so the transport cost will be 6 tonne-km (i.e. 3 kilometres × 2 tonnes). In Figure 2.9(a) the isotims, which refer only to the cost of transporting the raw material, are circular and equidistant. This shows that the transport costs are the same in all directions from the raw material site.

Figure 2.9(b) shows the cost of transporting 1 tonne of the product to the market from any part of the area. These costs are much less than those for transporting the raw material because only 1 tonne of product is derived from 2 tonnes of raw material. The isotims show that transport costs increase radially from the market, but only at a relatively slow rate. The graph in Figure 2.9(b) is much flatter than that in Figure 2.9(a).

Figure 2.9(c) shows the transport cost isotims for both the raw material (Fig. 2.9(a)) and the product (Fig. 2.9(b)). At each intersection of the two sets of isotims the numbers represent the total cost (in tonne-km) of transporting the raw material from the raw material source and the product to the market. At the raw material source (RM) the 2 tonnes of raw material have

MAPS

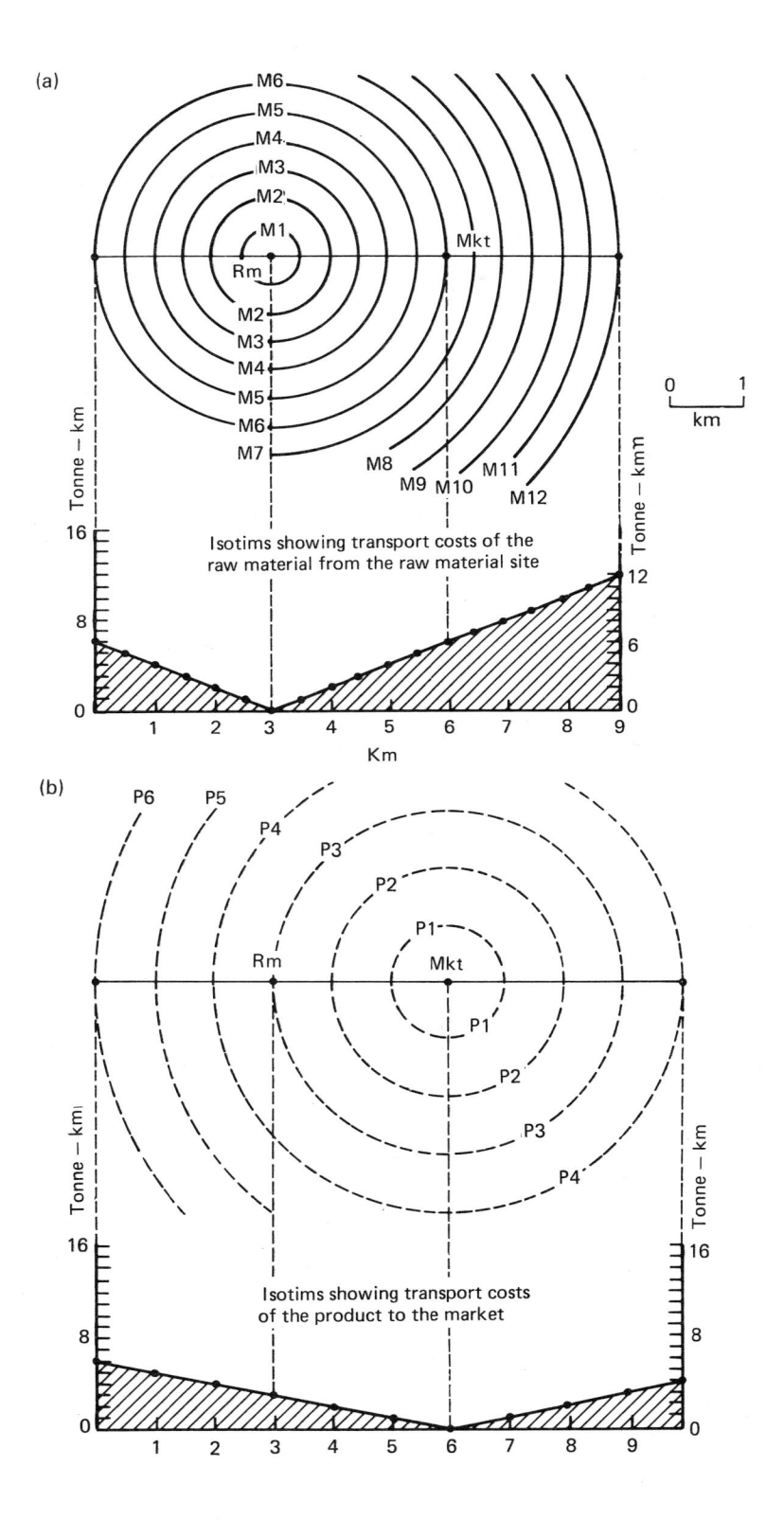

(a)

Isotims showing transport costs of the raw material from the raw material site

(b)

Isotims showing transport costs of the product to the market

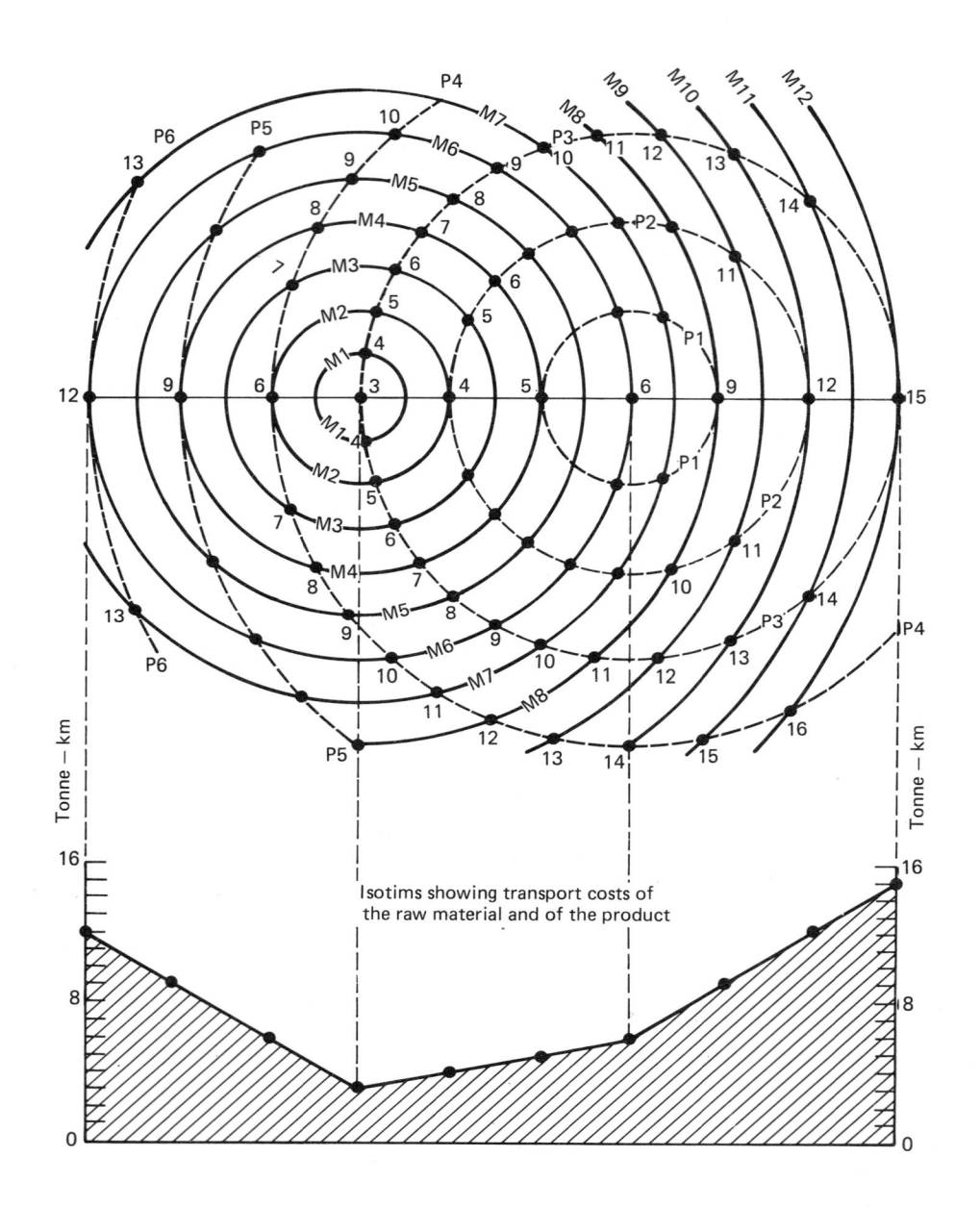

48 PRACTICAL GEOGRAPHY: PRESENTATION AND ANALYSIS

(c)

Isotims showing transport costs of the raw material and of the product

MAPS

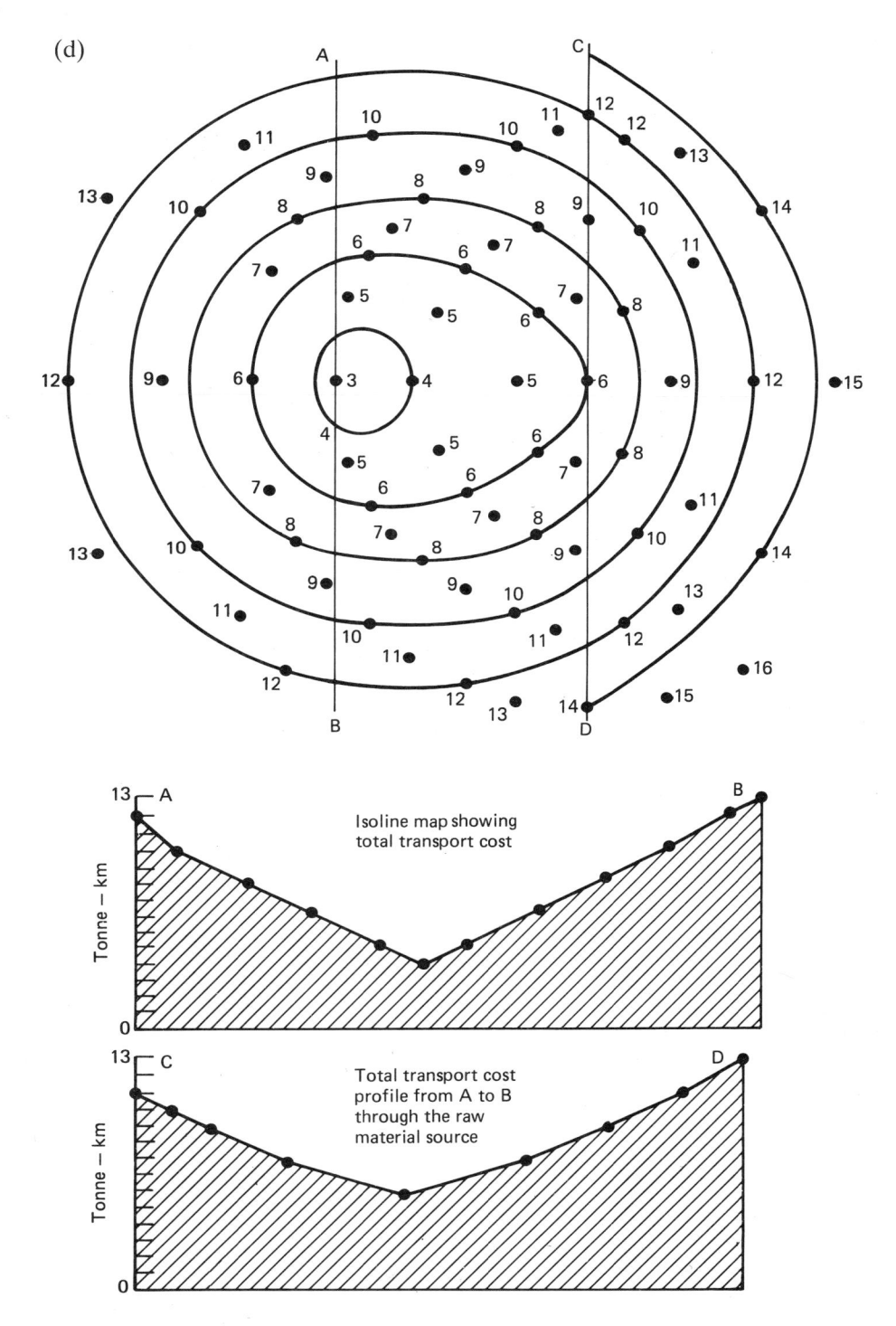

Fig. 2.9 Isotims and isodapanes

no transport cost because they are already there. The only transport cost is for the transport of 1 tonne of the product to the market, a distance of 3 km. Hence the total transport cost here is only 3 tonne-km. The market location has higher transport costs (6 tonne-km), all this being the cost of bringing 2 tonnes of the raw material a distance of 3 km. The highest value is 16 tonne-km (ringed on Figure 2.9(c)). This is made up of 12 tonne-km (2 tonnes × 6 km) for the raw material and 4 tonne-km (1 tonne × 4 km) for delivering the product to the market.

At each intersection of the isotims in Figure 2.9(c) the number indicates the total transport cost of obtaining sufficient raw material and then delivering 1 tonne of product to the market. These numbers have been plotted in their same relative positions in Figure 2.9(d). Isolines of total transport cost (isodapanes) have then been sketched in relation to the point values. Thus it is easy to see the general spatial pattern of transport costs. The lowest transport costs of all (under 4 tonne-km) are near the raw material. From here they rise quite steeply to the 'west'. To the 'east' however transport costs rise gradually as far as the market and then they steepen. The general impression is of a bowl shape slightly elongated from 'east' to 'west', its greatest depth being off-centre to the 'west'.

Other geographical applications
There are many other geographical applications of the isoline technique in addition to representing simple spatial variations in relief, temperature, rainfall, pressure and transport costs. Instead of simply representing the altitude of the ground surface or the average rainfall, temperature or pressure, it is possible to use isolines to portray the *variability* of relief and weather characteristics by mapping, for example, their standard deviation. Isolines can be drawn to represent any characteristic for which point values can be calculated and which is continuously distributed over space. It is possible for example, to use isolines to depict accessibility by road in an area. Using an Ordnance Survey map or a road map the total length of road within equally sized squares, or, preferably, hexagons can be measured. This length value is then recorded at the centre of each square or hexagon (Fig. 2.8(a)). Isolines can then be drawn at suitable intervals (Fig. 2.8(b)). In this case the variable is the density of roads. Isochrones are lines representing equal travel time from a given point. They can be constructed from information provided in a train or bus timetable or by timing a number of car or cycle journeys.

2.5 Flow diagrams

Flow diagrams are used to illustrate the various types of movement from place to place. They can illustrate such movements as the discharge of a river, flows of vehicles along roads and pedestrians along pavements, migrations from country to country, and shopping expeditions from villages to towns. There are two main types of flow diagram. One is the *flow line map* which illustrates the volume of movement or traffic that occurs along routes or channels that can be represented by a line or a network of lines such as roads, railways and rivers. The other is the *desire line map* which illustrates only the places of origin and destination of the movement and takes no account of the actual details of the route which is taken. A flow line map would be needed for example to show the amount of traffic along a road or railway. A desire line map is usually sufficient to show the pattern of international migrations of population since we are mainly interested in the sources and destinations. In a flow line map in contrast we are interested only in the routes which are used. We are not concerned with the original sources and ultimate destinations of the elements that are moving.

FLOW LINE MAPS

To construct a flow line map of the movement of traffic along roads it is necessary first of all to conduct a traffic count. Figure 2.10 shows the results of such a count which was taken on the northern outskirts of Bolton. The sites for such a count should be chosen with care.

Survey difficulties arise if the sites are placed too near busy road junctions and accurate recording is almost impossible where there are roundabouts or traffic lights. It is best to choose

MAPS

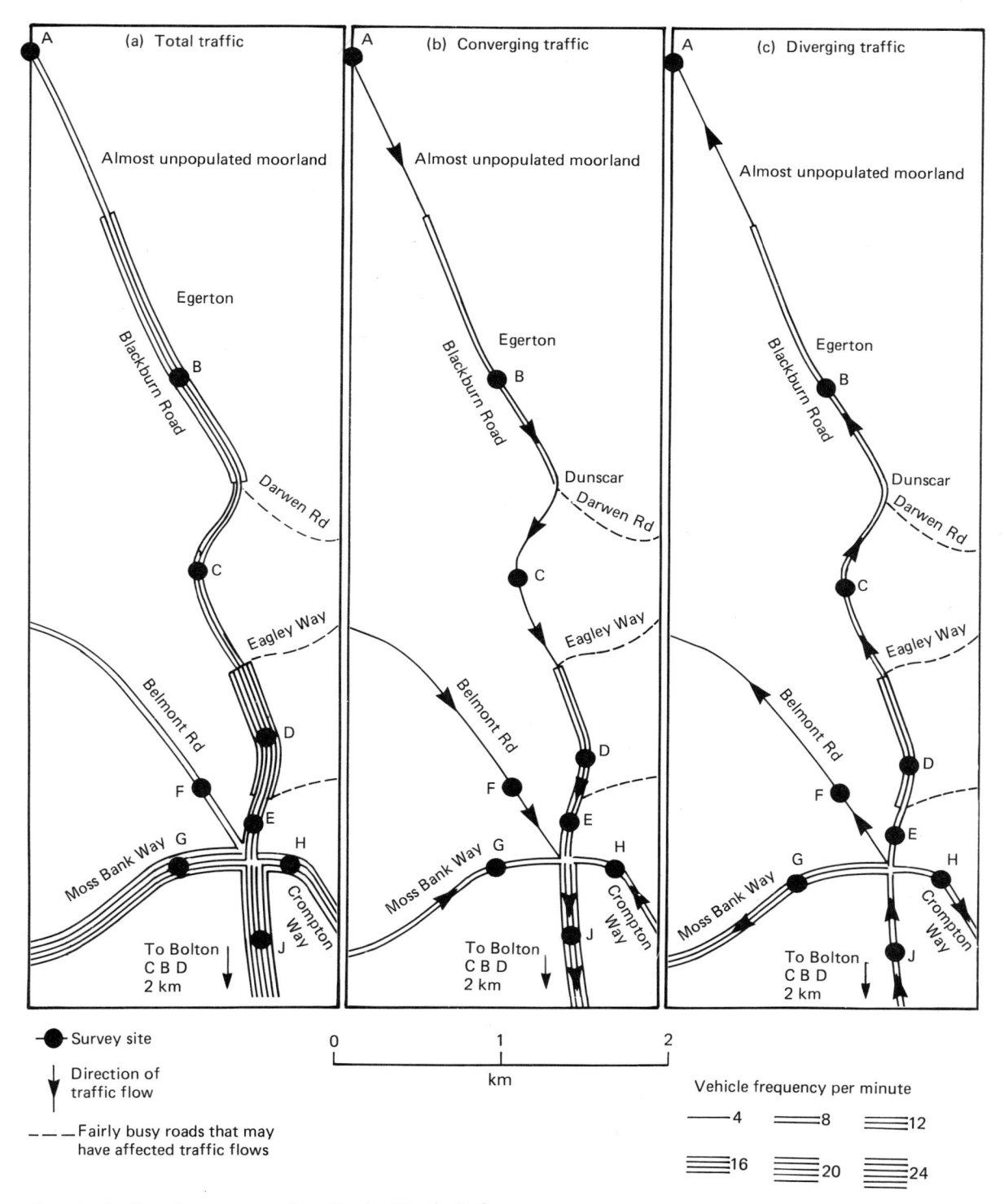

Fig. 2.10 Flow line maps of traffic in North Bolton

a site on a stretch of road where traffic can move fairly freely. The number of vehicles passing the survey site in each direction should be recorded separately. At each survey site a traffic count lasting five minutes will usually give a reasonable result and the survey should take place simultaneously at all the sites. It is also important to try to identify any other roads that may be influencing traffic flows, such as Darwen Road and Eagley Way in Figure 2.10.

The traffic count illustrated in Figure 2.10 was taken at 10am on a Friday morning at nine different locations. It shows the amount of traffic that passed along the major roads of north Bolton in a period of five minutes. Blackburn Road runs almost due north from central Bolton. Moss Bank Way and Crompton Way form a ring road in the shape of a semicircle with a radius of 2 km round the northern part of the town.

Figure 2.10(a) shows the total amount of traffic that passed along the main roads of north Bolton during the period of the survey. The amount of traffic passing along Blackburn Road generally increased from north to south, from about 8 vehicles per min to the north of Egerton (A) to about 24 vehicles per min to the south of the junction with the ring road (J). Traffic was also quite heavy along the two branches of the ring road (G and H) with about 16–20 vehicles per min. Belmont Road (F) with 8 vehicles per min was less busy. There was a relatively low traffic density at survey site C on Blackburn Road between the junctions with Darwen Road and Eagley Way (12 vehicles per min).

It is easier to understand the details of the traffic flow by studying Figure 2.10(b) and (c). Figure 2.10(b) (converging traffic) shows the volumes of traffic moving southwards towards the central business district (CBD). Traffic flows are generally greater than those shown in Figure 2.10(c) which are moving mainly northwards. The greatest single southward flow is at J where 16 vehicles per min are heading towards the CBD. This is twice as many as are moving northwards in Figure 2.10(c). To the north of the ring road however traffic flows northwards and southwards are generally equal. It is interesting to note that very little (4 per min) converging traffic passes survey site C (Fig. 2.10(b)). Traffic is heavier both to the north of the Darwen Road junction and to the south of the Eagley Way junction. It appears therefore that some southbound traffic from Egerton flows into Darwen Road and that, further south, a considerable amount of traffic joins Blackburn Road from Eagley Way. In Figure 2.10(c) it appears that some traffic is leaving Blackburn Road and turning into Eagley Way since the traffic flow along Blackburn Road is reduced from 12 per min to 8 per min north of the Eagley Way junction. Further north however the Darwen Road junction seems to have little effect on the northward flow through Egerton.

In Figure 2.10 the volume of traffic flows along the roads has been shown by allocating one line on the map to every four vehicles per min. This is a relatively simple, and quite effective, way of portraying traffic volumes along lines. Such a system could be used in a case in which material measured in hundreds or thousands of tonnes was moving from place to place. One line could represent a certain weight of material. It could also be used to illustrate the movements of trains along railways. Many flow line maps however indicate the volume of flow along a line by varying the width of *one* continuous line. For example, a width of 1 mm could be made to represent say, 10 vehicles or 100 tonnes weight. The line would widen in response to an increase in the amount of material moving. Sometimes the width of the line on the map bears a simple relationship to the values represented. In the table below, in Method 1, the value represented is divided by 10 and this gives the width of the line on the map in millimetres. In Method 2 the width of the line in millimetres is the square root of the value represented.

Value represented	Method 1 (divide value by 10)	Method 2 (square root of value)
1000	100 mm	31.6 mm
500	50 mm	22.4 mm
100	10 mm	10.0 mm
80	8 mm	8.9 mm
60	6 mm	7.7 mm
40	4 mm	6.3 mm
20	2 mm	4.5 mm

This table makes it clear that it is easier to represent a wide range of values by Method 2 than Method 1. In Method 2 a value of 1000 is represented by a line less than one-third of the width of the corresponding line in Method 1. On the other hand the line widths in Method 1 would be much easier to relate to their corresponding values. In Method 2, if the line width is doubled the value represented increases to much more than double. For example, in Method 2 a line width of 4.5 mm represents a value of 20, but a line width of 8.9 mm (approximately double) represents a value of 80.

DESIRE LINE MAPS

A *desire line* is a line on a map that represents the movement of people from their homes to certain destinations such as schools, workplaces, shopping centres, places of entertainment, holiday resorts or other types of central places on which flows of people converge. Peoples' homes are scattered over a considerable area and desire lines will tend to converge on these central places.

A desire line is drawn in a straight line from the home or the home area to the destination. It takes no account of the details of the actual route followed through a road, rail or air route network. Desire lines can be drawn for movements at any scale, from international migration and Mediterranean holidays to visits to the local village school, or from daily trips to the village school to periodic journeys to a university at the beginning of term.

Unlike a flow line the desire line shows only the locations of the origin and the destination of the journey and it does not always show variations in the volume of population movement. It is sometimes possible however to estimate the volume of movement from a particular locality to a particular destination by conducting a questionnaire survey either in the home district or, perhaps more conveniently, at the destination. For example, shoppers in a central business district may be asked to state the district in which they live. The desire lines can then be varied in width according to the number of people who have travelled from a particular area to a particular destination, as in the case of a flow line map.

In the case of journeys made for shopping in particular, but also for other services, a distinction is usually made between 'high order' and 'low order' services. Generally low order (e.g. 'first order') services are those which are demanded frequently and are available at a greater number of locations than higher order (e.g. 'second order' and 'third order') services which are demanded less frequently. The number of fish and chip shops in an area will almost always be greater than the number of stockbrokers' offices. This means that higher order service centres are usually further apart than lower order ones. Fish and chip shops are quite numerous in villages and small towns as well as large towns, but stockbrokers' offices tend to be confined to sizeable towns. Also, in a village it is usual to find a greengrocer, a hardware shop, a newspaper shop, an inn and perhaps a small branch bank. Such a village may be classed as a first order central place. A larger village might have a greater number and variety of shops, a larger bank and perhaps a few offices such as a building society and a small insurance office. This may be classified as a second order central place. Higher order central places than villages would have more, and a greater variety, of central functions, such as large department stores, several banks and building society offices and a greater variety of shops. Thus, if a resident in a village requires services of higher order than are available in his village he will have to travel to a higher order service centre.

In Figure 2.11 desire lines have been used to illustrate patterns of movement in relation to three orders of central place. Figure 2.11(*a*) shows the movements that are necessary to obtain first order services. These are available at all nine service centres of the area. It is assumed therefore that each household obtains the service at its nearest central place. The result is that journeys are very short, with an average length of less than 2 km, and only one journey is longer than 3 km. Figure 2.11(*b*) shows the pattern of desire lines associated with movement to obtain second order services. These are provided by only five second and third order central places. The result is that journeys in some cases are rather longer. One is over 4 km in length and the average distance between home and service centre is over 2 km. Finally Figure 2.11(*c*) shows the pattern of movement to obtain third order services. In this area only one central place provides this level of service, so all households have to travel to the same central place. The average travel distance is now over 4 km.

54   *PRACTICAL GEOGRAPHY: PRESENTATION AND ANALYSIS*

(a) Movement to first order centres

(b) Movement to second order centres

(c) Movement to the third order centre

Location of home **x**

Location of service centre (central place)

    First order   ●

    Second order   ◉

    Third order   ⊙

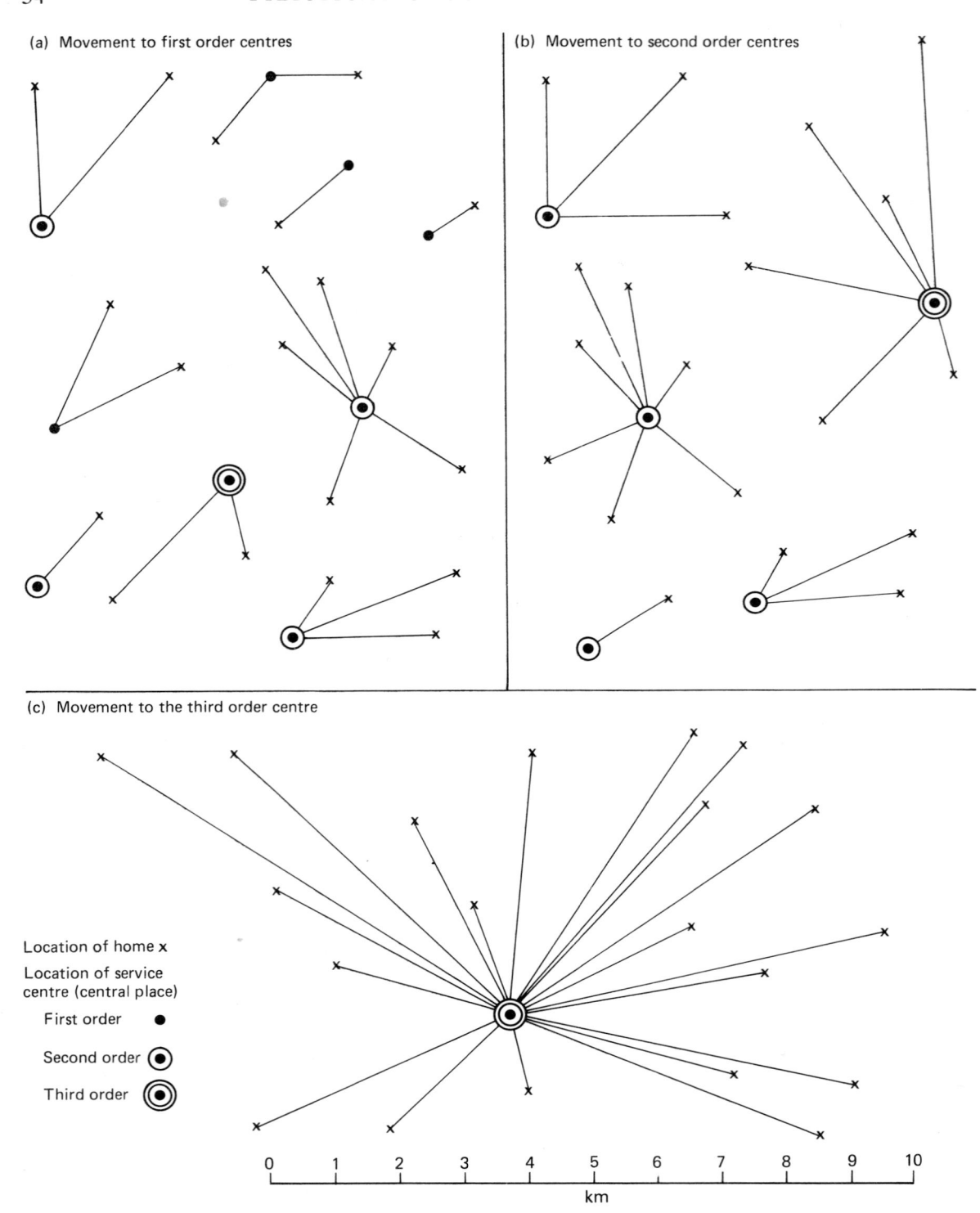

0 1 2 3 4 5 6 7 8 9 10
km

Fig. 2.11 Desire lines

Exercises

1. The map below shows some of the countries in northern Africa which have very striking contrasts in population density. Using the information given below construct a dot map of the distribution of population.

   It is recommended that each dot should represent 1 million people and that each dot should be small (46 dots have to fit into Egypt). Reference to an atlas map would also be of value so as to identify areas with a desert climate.

   It is also possible to trace the above outline and compile a choropleth map on the basis of the population density values (per km^2) given below:

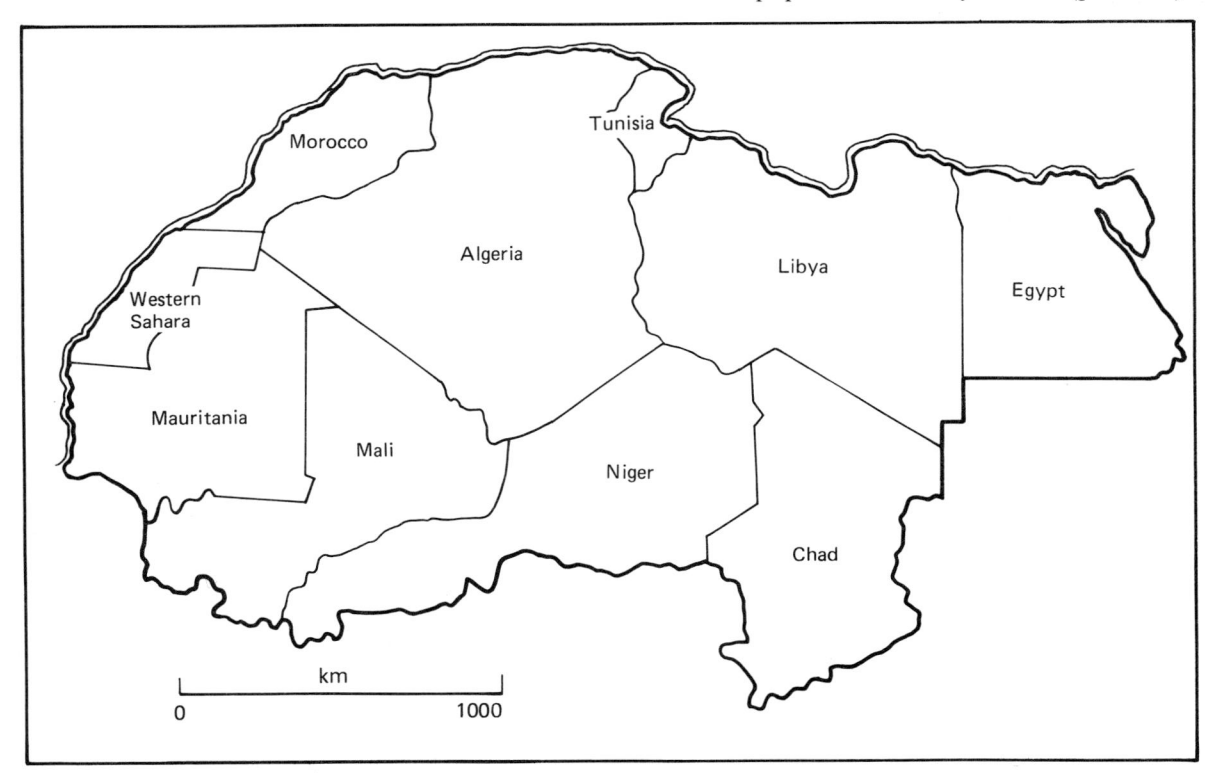

   Morocco 48, Algeria 9, Libya 2, Egypt 45, Chad 4, Niger 5, W Sahara 0.6, Mauritania 2, Mali 6, Tunisia 41.

2. Referring to the information given in question 1, trace two outlines of the political boundaries.
   (a) On one of these outlines use proportional bars rising vertically to represent the population totals of the different countries.
   (b) On the other outline map use proportional circles to represent the population totals.

	Population of country (millions)	Million cities
Morocco	23	Casablanca
Algeria	21	Algiers
Libya	3.6	
Egypt	46	Cairo(5m), Alexandria(2m), Giza
Chad	5	
Niger	6	
W Sahara	0.3	
Mauritania	2	
Mali	7	
Tunisia	7	Tunis

(c) Comment on the advantages and disadvantages of using dots, the choropleth method and proportional bars and circles to represent the distribution of population on maps of this scale.

3. In each of the following cases, state the location(s) at which the raw material is processed and the product delivered to the market with the least possible transport cost. Transport cost is measured in tonne-km. In these exercises you should assume that movement is equally easy in all directions. In each case illustrate your answer with a diagram.
    (a) The raw material is situated 10 km from the market for the product. 1 tonne of raw material is processed into 1 tonne of product.
    (b) The raw material is situated 15 km from the market for the product. 3 tonnes of raw material are processed into 1 tonne of product.

4. Draw a diagram to show the following features.
    (a) Insert isotims to show both raw material and product transport costs. The raw material transport costs should be lower than those for the product.
    (b) Using the pattern of isotims you have drawn insert a set of isodapanes.

(c) Indicate the location at which total transport costs are least and describe their variations over the area.

5. Make a copy of the map below showing the distribution of a number of values of a spatially continuous variable. On your map draw isolines for values of 10, 20, 30, 40, 50 and 60 units. It is best to begin by plotting the isolines for 10 units and 20 units.

63	58	62	65	68	72	75
56	52	50	56	60	64	66
60	59	50	40	49	55	62
62	58	55	47	30	40	50
56	49	52	52	32	24	40
49	35	37	45	30	14	26
37	26	20	26	20	8	20

3 Matrices

A matrix is a square or rectangular pattern of cells or compartments (Fig. 3.1). Each cell often contains a numerical value. The lines of cells running horizontally across the matrix are called *rows*; those running vertically are called *columns*. Figure 3.1(*a*) is a simple example of a matrix. It shows the distance in kilometres between each pair of the locations A, B, C, D and E. For example, from location A to location E is 8 km and from location D to location C the distance is 4 km. Because the matrix has 5 rows and 5 columns it is described as a 5×5 matrix. Because there are equal numbers of rows and columns it is described as a *square* matrix. The *principal diagonal* runs across the matrix from top left to bottom right and it is easy to see that the same cell values occur, as if reflected, on both sides of this line. A matrix with this characteristic is said to be *symmetric*.

In such a matrix each point to point distance is given twice. The distance of 2 km between A and B is indicated both on row A at column B and on row B at column A. Similarly every other distance is indicated twice. It is possible therefore to omit half of the matrix without any loss of information. Figure 3.2(*b*), with only 10 cells, contains just as much information as (*a*), which has 20 cells. The vertical column of successive distances from location C in (*a*) is 4, 3, 4, 6. In (*b*) the same distances are shown first vertically for the 4 and the 3, then horizontally for the 4 and the 6.

Both (*a*) and (*b*) illustrate the relationships between the members of a *single* set of locations. The same locations, A, B, C, D and E, occupy both the rows and the columns. It is therefore possible to calculate the distance between each location and all the remaining locations taken together. This is done by adding the total value for either each of the rows or each of the columns. Location A, for example, is at a distance of 20 km from B, C, D and E taken together (2 km from B, 4 km from C, 6 km from D and 8 km from E). B's total distance is 17 km (2 + 3 + 5 + 7). C's total distance is also 17 km, D's is 20 km and E's is 26 km.

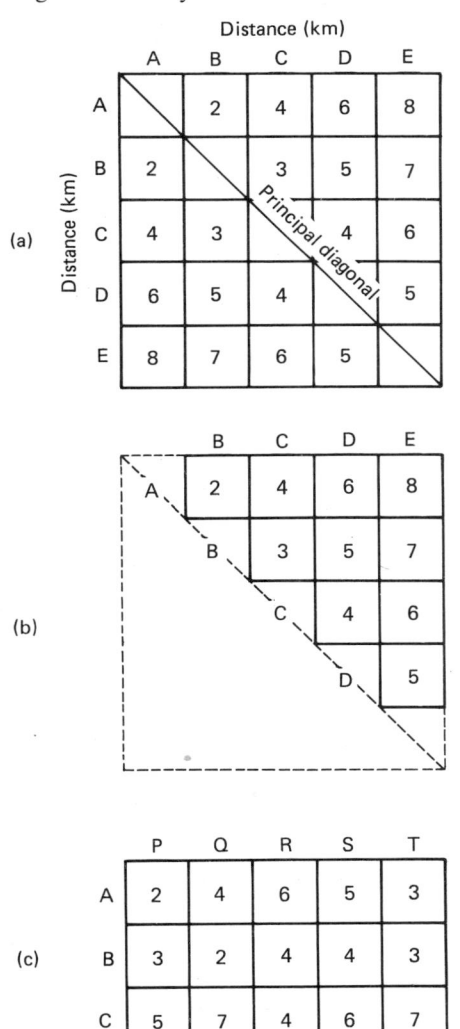

Fig. 3.1 Matrices

Climatic Statistics

	Mean temperature (°C)		Mean precipitation (mm)	
Location	January	July	April–September	October–March
London	4	7	300	325
Hamburg	0	17	381	318
Vienna	−2	20	378	246
Lisbon	10	21	193	560
Bergen	1	14	875	1168
Barcelona	8	23	260	277
Athens	9	27	99	302

Matrix

	Mean temperature (0°C)		Mean precipitation (mm)	
	January under 8°C	July over 20°C	April–Sept over 500mm	Oct–March over 500mm
London	1	0	0	0
Hamburg	1	0	0	0
Vienna	1	0	0	0
Lisbon	0	1	0	1
Bergen	1	0	1	1
Barcelona	0	1	0	0
Athens	0	1	0	0

Classification

Climatic type (from matrix)	Description	Examples
1 0 0 0	Fairly cold winter; cool summer; precipitation less than 500mm in both summer and winter	London, Hamburg, Vienna
0 1 0 1	Fairly warm winter; hot summer; fairly dry summer; fairly wet winter	Lisbon
1 0 1 1	Rather cold in winter and cool in summer; precipitation greater than 500mm in both summer and winter	Bergen
0 1 0 0	Fairly warm winter; hot summer; Fairly dry in both summer and winter	Barcelona, Athens

Fig. 3.2 Classification of climates by means of a matrix

The information provided in Figure 3.1(c) is rather different. In the first place the matrix has only three rows and is therefore described as a 3 × 5 matrix. Also it shows the relationships between one set of locations, A, B and C and another, quite different, set, P, Q, R, S, and T. We can deduce that location A is 20 km distant (2 + 4 + 6 + 5 + 3) from P, Q, R, S and T taken together, but we can learn nothing about the distances between A and B or B and C or P and Q. We are dealing here with two, quite different, sets of elements.

Classifying data

A matrix can also be used to classify data. In Figure 3.2 climatic statistics consisting of four separate numerical values (mean temperature for January and July and mean precipitation for April to September and October to March) are listed for seven places in Europe. From the table it is not easy to recognise overall patterns because the numerical values differ so greatly. In the matrix two specimen values for temperature (January under 8°C and July over 20°C) and two values for precipitation (over 500 mm in summer and over 500 mm in winter) are given at the heads of the columns. The locations of the seven places are given at the ends of the rows. The actual temperature and precipitation values do not appear in the matrix. Instead a 1 is inserted if a particular location possesses the climatic characteristic stated at the top of the column and a 0 is inserted if it does not. It can now be seen that the climates of London, Hamburg and Vienna, all of which have rows consisting of 1 0 0 0, are generally fairly similar. Also Barcelona is somewhat similar to Athens, both having 0 1 0 0 in their row. Lisbon differs from these in respect of its heavier winter precipitation.

The same technique (1 representing 'yes' and 0 representing 'no') can be used to study the relative location of places. In Figure 3.3 a map shows the location of the boundaries between eleven countries in Western Europe. We wish to discover which of these countries is the most 'central'. The matrix has been constructed by inserting a 1 in each cell that represents a pair of countries with a common boundary and a 0 for cells representing a pair with no common boundary. The same countries have been listed at the ends of the rows and at the heads of the columns of the matrix, so the matrix is square and symmetric. Hence it is only necessary to show half of the matrix (Fig. 3.3). This prevents 'double scoring' and the number of contiguous (neighbouring) countries is easily counted in the

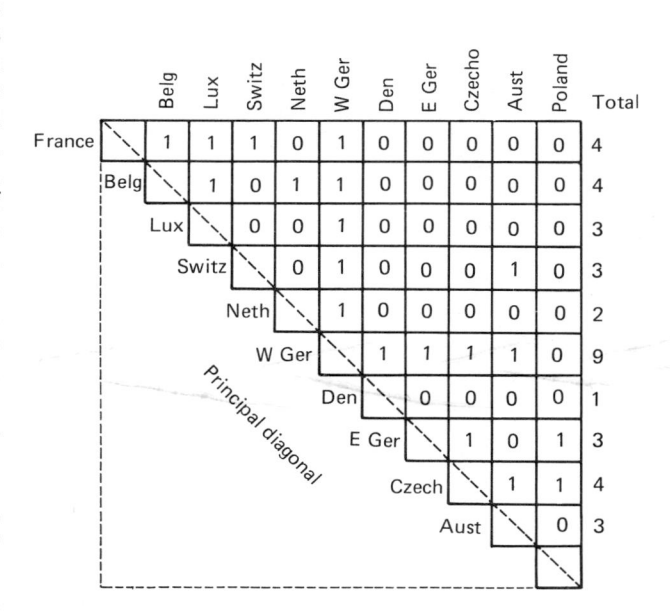

	Belg	Lux	Switz	Neth	W Ger	Den	E Ger	Czecho	Aust	Poland	Total
France	1	1	1	0	1	0	0	0	0	0	4
Belg		1	0	1	1	0	0	0	0	0	4
Lux			0	0	1	0	0	0	0	0	3
Switz				0	1	0	0	0	1	0	3
Neth					1	0	0	0	0	0	2
W Ger						1	1	1	1	0	9
Den							0	0	0	0	1
E Ger								1	0	1	3
Czech									1	1	4
Aust										0	3

Fig. 3.3 Contiguity of countries in Western Europe

matrix by starting at the head of the column and 'reflecting' to the right at the principal diagonal. There are eleven countries on the map, so the maximum number of common boundaries that a country can have is ten. It cannot have a common boundary with itself. West Germany has the greatest number of common boundaries, being linked to all the other countries except Poland. Denmark has fewest, being linked only with West Germany. Matrices can be used in many other types of geographical study. An obvious example is in displaying the flows of imports and exports between countries. They can also be used to show flows that take place between the industries or the wider economic sectors of a country in the form of input–output matrices. Their importance is not restricted to the representation of flows. Connectivity matrices are used in network analysis. Indices showing the amount of interaction that takes place between towns or areas can be displayed usefully in the form of a matrix. It is also possible to construct a matrix which displays similarities and contrasts between regions or countries. In this case the values shown in the matrix can be correlation coefficients.

	A	B	C	D	E
A					
B					
C					
D					
E					

2. Construct a matrix, similar to that in Figure 3.3, showing the contiguity of the North African countries shown on the map on page 55.

Exercises

1. The diagram below shows a road network in which only one-way movement is possible along two of the links. Construct a matrix showing the distance between each pair of locations in each direction. Your matrix should be set out as shown.

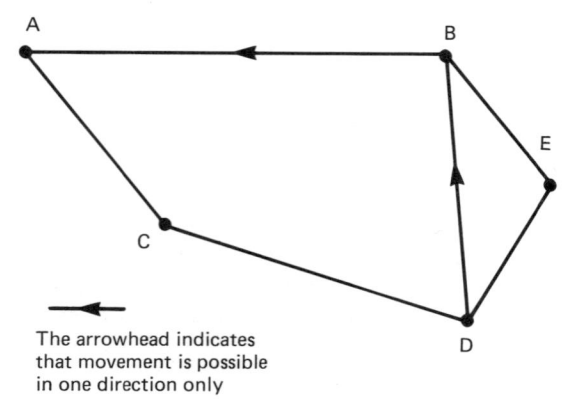

The arrowhead indicates that movement is possible in one direction only

4 Techniques Used in the Analysis of Data

4.1 Averages

One of the simplest procedures that can be applied to geographical data is to calculate some kind of average which summarises the general characteristics of the data. Ideally such an average should be determined in the light of all the values that are included in the data. The average value should also be 'typical' of the data and therefore should resemble values that occur centrally within the range of the data. These points are important because the average will be used to make comparisons with other sets of data, as in the case, for example of values that are quoted for mean annual precipitation or temperature. The average is likely to be more accurate if it has been derived from a set of individual values that are clustered within a narrow range. If the values in the data are widely dispersed the average derived from them will be less reliable. The main averages used in geography are the *mean* (sometimes called the average) the *median* and the *mode*.

The mean

The *mean* (arithmetic mean, $\bar{x}$) is calculated by adding together all the values of a variable (Σx), then dividing this total by the frequency (n) which is the number of separate values there are in the data. This process is described mathematically by:

$$\bar{x} = \frac{\Sigma x}{n} \quad \text{(sum of values)} \atop \text{(number of values)}$$

The mean is useful because it summarises a large amount of numerical data and thus makes it easy to compare and classify two or more sets of data. One can compare, for example, the mean temperatures or the mean precipitation values of the United States with those of Europe or Africa by calculating their mean values. Unlike the median and the mode, the other 'measures of central tendency', all the values of the variable contribute to the calculation of the mean.

Besides calculating the mean by using a complete list of separate values in the data, the mean can be calculated from a histogram (see page 2) in which no individual values of the variable are depicted. When using a histogram the number of occurrences in any value class are assumed to be distributed equally around the midpoint of the class (Fig. 1.2). Thus, for every class, the frequency is multiplied by the value of the class midpoint. This gives a total value for each class. The mean is then obtained by adding the values of all the classes and then dividing by the total number of frequencies. This process is described by:

$$\bar{x} = \frac{\Sigma \, (\text{frequency} \times \text{class midpoint})}{\Sigma \, \text{frequencies}}$$

A characteristic of the arithmetic mean is that the sum of the positive and negative deviations from the mean is zero. The mean value of 10 4 6 8 5 and 3 is 36/6, i.e. 6. If we now subtract 6 from each of these values we have:

10	4	6	8	5	3	
−6	−6	−6	−6	−6	−6	
4	−2	0	2	−1	−3	= 0

The mean is particularly useful for summarising a set of data when the values cluster around a central value but much depends on the kurtosis of the histogram (page 4). In Figure 4.1(*a*) the histogram is fairly leptokurtic, with the rectangles rising markedly in the centre. The mean value of the variable coincides with the peak frequencies, so that a large number (16) of values of the variable are similar to that of the mean. In Figure 4.1(*b*) the central peak no

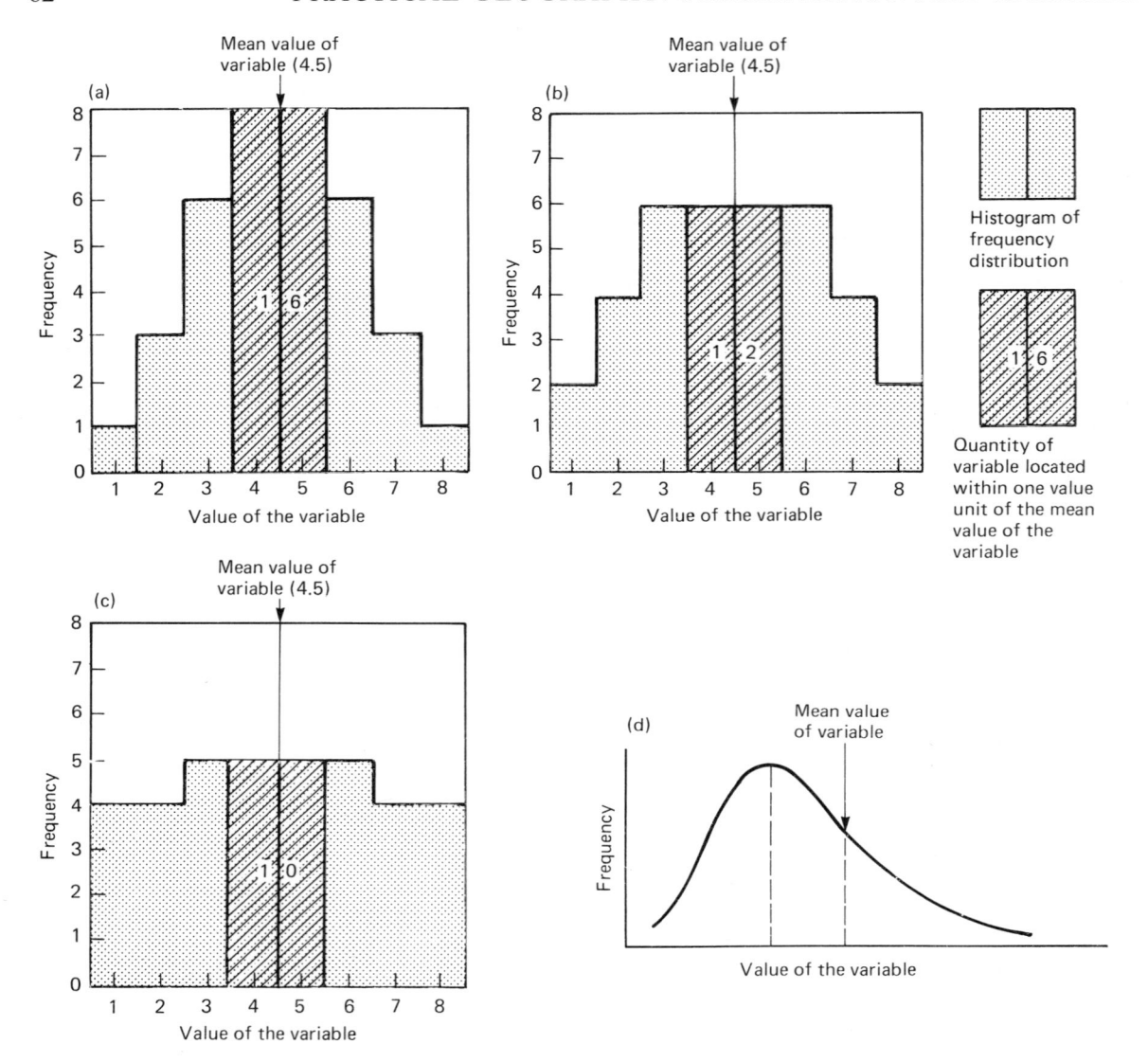

Fig. 4.1 The arithmetic mean

longer exists, so there are fewer (12) values similar to the mean. In Figure 4.1(c) the histogram is distinctly platykurtic and fewer still (10) values are similar to the mean. So the usefulness of the mean in summarising data is related to the kurtosis of the data. It is also related to the skewness of the data. The mean is less likely to provide a good summary of the data set if the histogram is skewed (page 4). In Figure 4.1(d) the histogram is positively skewed and this has displaced the mean away from the values with the highest frequency.

A useful adaptation of the arithmetic mean is the *running mean* or *moving average* (page 13, Fig. 1.14), which smooths the fluctuations that occur through time and identifies general trends.

The median
The *median* of a set of numbers arranged in order of size is the middle value of the set. If the numbers are set out in order of magnitude,

TECHNIQUES USED IN THE ANALYSIS OF DATA

as in an array, the median is the $\frac{n+1}{2}$ th value. To find the median therefore we count the number of values, add 1 and then divide by 2. In the following set of numbers there are 10 values:

1 2 2 2 3 5 7 7 8 9

The median value is the $\frac{10+1}{2}$ th (i.e. 5.5th) value from either end of the array. This is midway between the two central values of 3 and 5, so the median is 4 even though there is no value of 4 in the data. With an even number of values the median is always the mean of the two central values. If there is an odd number of values the same procedure is used. In the following set of numbers there are 9 values:

2 2 2 3 6 7 7 8 9

The median in this case is the $\frac{9+1}{2}$ th (i.e. 5th) value from either end. In this case the value is 6.

It is also possible to make an estimate of the value of the median from the information provided on a histogram, but because in a histogram the values of the variable have been grouped into classes it is not possible to calculate a completely accurate value for the median. In the histogram in Figure 4.2(a), for example, we can identify the central item of the variable (the median) by counting the total frequency (2+3+6+4+3+3+2+2 = 25). 25+1 = 26, and 26 divided by 2 gives the median as the 13th value from either end of the histogram. The class in which this middle value occurs is known as the *median class*. In this distribution there are 11 items below the median class (in classes 1, 2 and 3) and 10 items above the median class (in classes 5, 6, 7 and 8). We can therefore estimate that the median value occurs in the 4th class slightly nearer to its lower class boundary. This suggests that the median is slightly under 4 value units of the variable. The median value is greater than that represented by the largest rectangle of the histogram.

It is also possible to read off the value of the median from a cumulative frequency graph (ogive) (Fig. 1.4).

An important characteristic of the median is that its mean deviation from the other values

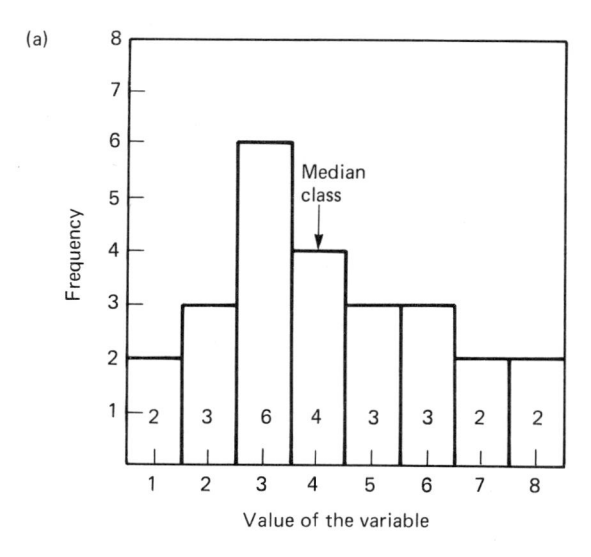

(a)

(b) Deviations from the values of a distributions

	Values of the distribution								Total deviation	Mean deviation
	2	2	2	3	6	7	7	8	9		
2	0	0	0	1	4	5	5	6	7	28	3.1
3	1	1	1	0	3	4	4	5	6	25	2.8
→6 (median)	4	4	4	3	0	1	1	2	3	22	2.4
7	5	5	5	4	1	0	0	1	2	23	2.6
8	6	6	6	5	2	1	1	0	1	28	3.1
9	7	7	7	6	3	2	2	1	0	35	3.9

Fig. 4.2 The median

of the variable is less than at any other location. Thus it provides a valuable generalisation about the data. Figure 4.2(b) illustrates the variations in the total and the mean deviation that can occur between the values of a distribution. The deviations among the central values of the distribution (particularly 6—the median) are considerably smaller than those related to the extreme values of 2, 8 and 9.

The median therefore is a useful measure of central tendency. It can even be used in cases where numerical data is incomplete. In that case it is possible to include in the data categories such as '100 and over' and 'less than 10' provided that the number of values is known. Also, the median depends only on one or two specific values at the centre of the distribution. Unlike the mean, the median is an actual value in the distribution unless it is the average of two values in the centre of an even-number of values. It is much easier to calculate than the mean but, unlike the mean it cannot be used for further statistical processing.

The mode

In a frequency distribution the single value that occurs with the greatest frequency is called the *mode*. It is the most popular or fashionable or typical value. Like the median it has the advantage of not being influenced by extreme values at each end of the distribution unless they become quite large. A distribution with a single peak is referred to as unimodal. If there are two peaks it is described as bimodal. In a histogram the largest rectangle represents the modal class, which has most members. Some variables are perhaps best described by using the mode. These could include the sizes of men's shoes, and population densities in rural areas. In a histogram representing continuous data the largest rectangle indicates the modal class. The value of the mode can be deduced from this by drawing lines as shown in Figure 4.3.

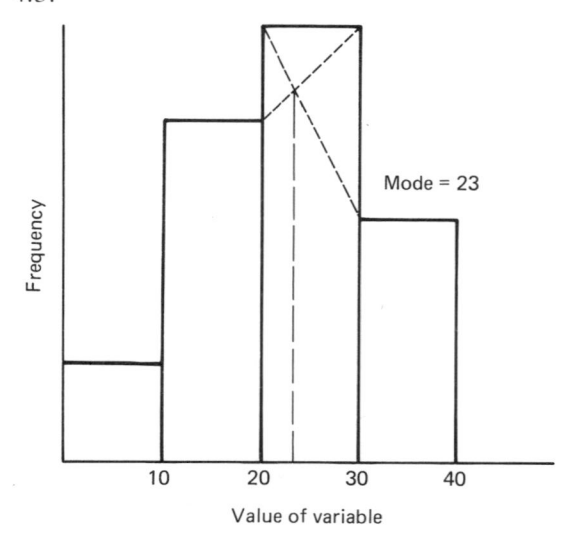

Fig. 4.3 Identifying the mode from a histogram

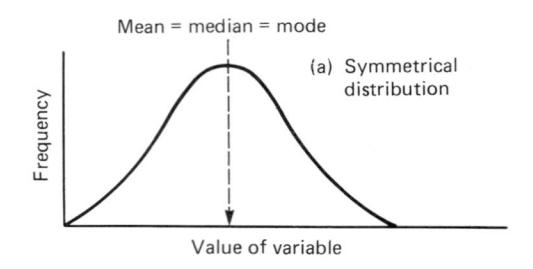

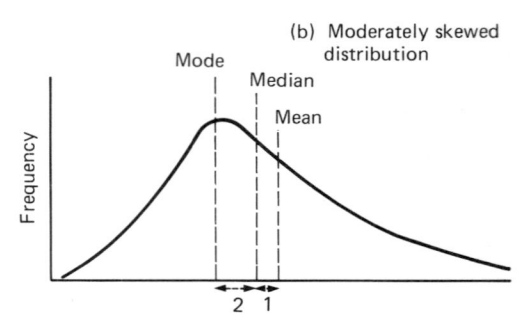

Fig. 4.4 Relationship between the mean, the median and the mode

Figure 4.4 shows the approximate relationship between the values of the mean, the median and the mode in a frequency distribution. If the distribution is symmetrical, as in Figure 4.4(*a*) the highest point on the curve must represent the mode and also the median and the mean. If, on the other hand, the distribution is moderately skewed, as in Figure 4.4(*b*), the highest point of the curve is still the mode, but the mean and the median have moved to the right. This is because extra frequencies have been added to the right of the mode to produce the skew. The mean has moved even further to the right because the extra frequencies added are of higher values of the variable. In the skewed distribution the approximate relationship between the mean, the median and the mode is that the difference in value between the mode and the median is twice that between the median and the mean. This is usually described by the relationship:

Mean − Mode = 3(Mean − Median).

If the skew in Figure 4.4(*b*) had been negative instead of positive the relationship between mean, median and mode would have been the mirror image of that shown.

4.2 Measures of dispersion

Frequency distributions differ widely in their shape. Many are approximately symmetrical but their height (frequency) and width (values of the variable) can both vary. In some cases many values of the variable are concentrated near the mean, thus giving a narrow but high (leptokurtic) peak frquency. In other cases the range of values is greater and the centre of the distribution rises to form a shallow dome. The various degrees of dispersion and the resultant shapes of the distribution can be described by the use of numerical indices.

The range

The simplest of these indices is the *range*. It is extremely easy both to calculate and to understand. The range, quite simply, is the difference between the highest value and the lowest value in the data. It is rather a weak index because it only indicates the difference between the two extreme values of the distribution. Two distributions are given below:

1 2 3 3 3 3 4 5
1 1 1 1 1 1 1 5

In both of these cases the range is 4 (i.e. $5 - 1$), yet the details are very different. Also, it seems unlikely that either the highest or the lowest value could have much relationship with the remaining values. Figure 4.5 shows three different frequency distributions all of which have the same range. The weakness is that the range gives no information at all about the absolute values of the variable.

Quartiles

The *quartiles* are the items that divide the array of values into four equal parts. They are best illustrated by relating them to the simple array of values of the following variable:

1 1 2 ③ 3 4 5 ⑤ 6 6 6 ⑧ 8 9 9

There are 15 separate values in this array. The *first* (lower) quartile is defined as the $\frac{n+1}{4}$ th item from the lower end of the array. There are 15 items, so $15 + 1 = 16$, which, divided by 4, gives 4. So the fourth item (3) has been ringed in the array. The value of the first quartile is 3. The second quartile is the median. This is the $\frac{n+1}{2}$ th item from the lower end (the eighth) which has a value of 5. Finally, the third quartile is identified by the expression $\frac{n+1}{4} \times 3$ which makes it the twelfth item from the lower end, with a value of 8. The measure of dispersion that uses the quartiles is called the *quartile deviation* (or semi-interquartile range). It is calculated by subtracting the value of the first quartile from that of the third quartile and then dividing by 2. In this case the quartile deviation is $8 - 3 = 5$, which divided by 2 gives 2.5. So it follows that half of all the items differ from the median by not more than 2.5 units of value, i.e. they lie between 2.5 and 7.5.

The quartile deviation therefore is quite a useful measure of the dispersion of a distribution. At least it gives a good description of the half of the distribution that occurs between the upper and lower quartiles. Unlike the range, however, it gives no information at all about the distribution of values less than the lower quartile (Q_1) and greater than the upper quartile (Q_3). An advantage of this is that the quartile deviation cannot be disturbed by the accidental occurrence of a few 'freak' values at each end of the distribution. Another advantage is that the quartile deviation, unlike the range, can still be calculated if some of the items at each end of the distribution are classified simply as 'under 10' or 'over 50' and no precise values are available. As long as the frequency of such values is given the quartile deviation can be calculated but the range cannot.

Calculating the mean deviation

When calculating the *mean deviation* it is necessary first of all to find the arithmetic mean of the values in the distribution (page 61). Then the differences between the mean and each value in the distribution are calculated. These differences are positive in sign ($+$) because they are all regarded as positive deviations from the mean value. These differences (deviations) are then summed and the total is divided by the total *number* of values in the distribution. These processes are summarised in the relationship:

Mean deviation $= \dfrac{\Sigma (x-\bar{x})}{n}$.

An example is given below:

Values of the variable:
2 2 3 5 6 7 7 8 10 10

The sum of these values is 60 and there are 10 separate values, so the value of arithmetic mean is 6. We now subtract 6 from each value in the distribution:

$$\begin{array}{cccccccccc} 2 & 2 & 3 & 5 & 6 & 7 & 7 & 8 & 10 & 10 \\ -6 & -6 & -6 & -6 & -6 & -6 & -6 & -6 & -6 & -6 \\ \hline -4 & -4 & -3 & -1 & +0 & +1 & +1 & +2 & +4 & +4 \end{array} = 0$$

If we observe the + and − signs the sum of the differences is zero. In calculating the mean deviation however we ignore the + and − signs and consider only *deviations* so we find that the sum of the deviations is 24. The mean deviation therefore is 24 divided by 10, which is 2.4.

Usually the mean deviation is calculated in relation to the arithmetic mean but it can also be calculated in relation to the median or the mode (pages 62 and 64). The value of the mean deviation is the minimum possible when it is calculated in relation to the median (page 62). The mean deviation is easy to calculate and it takes into account all the items in the distribution but it is less often used than the standard deviation. The most commonly used index to describe the extent to which the values in a distribution tend to cluster around the mean is the *standard deviation*. This is sometimes referred to as the 'root-mean-square deviation' because its value is the *square root* of the *mean value* of the *squared deviations* from the mean value of the distribution.

Calculating the standard deviation

The formula for calculating the standard deviation is:

Standard deviation $= \sqrt{\dfrac{\Sigma(x-\bar{x})^2}{n}}$

It is calculated in the following stages:

(i) Calculate the arithmetic mean of all the values in the distribution ($\bar{x}$);

(ii) Calculate the deviation of each value from the mean $(x - \bar{x})$;

(iii) Square the deviation of each value from the mean $(x - \bar{x})^2$;

(iv) Add together all the squared deviations from the mean $\Sigma(x-\bar{x})^2$;

(v) Divide this last value by the number of items in the distribution $\dfrac{\Sigma(x-\bar{x})^2}{n}$. This value is referred to as the *variance* of the distribution.

(vi) Find the square root of the variance $\sqrt{\dfrac{\Sigma(x-\bar{x})^2}{n}}$ As an illustration we shall now calculate the standard deviation of the set of values that was used to demonstrate the calculation of the mean deviation above:

2 2 3 5 6 7 7 8 10 10

The sum of the ten values is 60, so the arithmetic mean is 6. Next we calculate the deviations from the arithmetic mean:

$$\begin{array}{cccccccccc} 2 & 2 & 3 & 5 & 6 & 7 & 7 & 8 & 10 & 10 \\ -6 & -6 & -6 & -6 & -6 & -6 & -6 & -6 & -6 & -6 \\ \hline -4 & -4 & -3 & -1 & 0 & 1 & 1 & 2 & 4 & 4 \end{array}$$

There is no need to ignore the minus signs because we now square each of the deviations and this removes the minus signs. We now add the squared deviations and divide by the number of values:

16+16+9+1+0+1+1+4+16+16 = 80
80 divided by 10 = 8.

The value of the *variance* is therefore 8. The square root of 8 is 2.83, so the *standard deviation is* 2.83. The mean deviation for the same set of values was 2.4 The mean deviation therefore for this set of values is slightly over 0.8 of the value of the standard deviation. For moderately skewed distributions this relationship is fairly general.

A different method is used to calculate the standard deviation of a grouped frequency distribution such as those shown in Figure 4.1, provided that the class intervals are equal. In this calculation it is assumed that all the values in any class are equal to that of the class midpoint. In the table below x represents the value of the class mid-point, f represents the class frequency and d represents the deviation (in whole classes) from an *assumed mean* which can be taken as the mid-point of *any one of the classes*. In Figure 4.1(*a*) the value of 3 has been selected. A table is drawn up with five columns headed respectively, x, f, d, fd and fd² (fd x d). A value of 3 has been selected as the assumed mean and has been allocated a deviation of zero. The table below refers to the histogram in Figure 4.1(*a*):

x	f	d	fd	fd²
1	1	−2	−2	4
2	3	−1	−3	3
3	6	0	0	0
4	8	1	8	8
5	8	2	16	32
6	6	3	18	54
7	3	4	12	48
8	1	5	5	25
Σ=36			Σfd=54	Σfd²=174

The column totals are then entered into the following equation:

$$\text{Standard deviation} = \sqrt{\frac{\Sigma fd^2}{\Sigma f} - \left(\frac{\Sigma fd}{\Sigma f}\right)^2} \times \text{class interval}$$

$$= \sqrt{\frac{174}{36} - \left(\frac{54}{36}\right)^2} \times 1$$

$$= \sqrt{4.83 - 1.5^2} \times 1$$

$$= \sqrt{4.83 - 2.25} \times 1$$

$$= \sqrt{2.58}$$

$$= 1.61$$

Calculated in a similar way the distribution illustrated in Figure 4.1(*b*) has a standard deviation of about 1.9 and that illustrated in Figure 4.1(*c*) has a standard deviation of about 2.2. One would expect this because in Figure 4.1(*a*) there are 16 items in the central two classes whereas, in (*b*) there are 12, and in (*c*) there are only 10.

The standard deviation is the best measure of the extent to which the values in a frequency distribution tend to cluster near the mean. In Figure 4.1 it was obvious that histogram (*a*) would have the lowest standard deviation. It has a frequency of 8 in each of the two central values near the mean and frequencies of only 1 in the extreme values. Histogram (*b*) has a frequency of only 6 in each of the central values, but has more than (*a*) in the extreme values. Histogram (*c*) with only 5 in each of the central values has a large number of frequencies in the extreme values. Thus, it is often possible to compare the standard deviations of distributions by studying the shape of their histograms or their frequency curves. For distributions that are only moderately skewed the standard deviation is about 1.25 times greater than the mean deviation and about 1.5 times greater than the quartile deviation. The standard deviation has the advantage of being calculated by the use of all the values in the distribution (like the mean deviation) but it is also the most useful measure of dispersion for further statistical processing. In a distribution which is symmetrical and unimodal about 95% of the items lie less than two standard deviations away from the mean and only about 1% lie more than three standard deviations away from the mean.

Calculating the coefficient of variation

Another measure of dispersion that is related to the standard deviation is the *coefficient of variation*. This is calculated by dividing the standard deviation by the arithmetic mean and then multiplying by 100:

$$\text{Coefficient of variation} = \frac{\text{Standard deviation}}{\text{Mean}} \times 100$$

In other words the standard deviation is expressed as a percentage of the mean. It is based on the principle that the variability of a frequency distribution depends on the value of both its mean and its standard deviation. A distribution with a mean of 40 and a standard deviation of 10 is relatively more variable than one with a mean of 80 and a standard deviation of 10. A deviation of 10 from 40 is a 25% deviation from the mean whereas a deviation of 10 from 80 is only a 14% deviation from the mean. It is possible for a distribution with a high standard deviation to be less variable than one with a lower standard deviation and a very low mean. The coefficient also allows us to compare the dispersion of distributions in which the mean and the standard deviation are expressed in different units. For example precipitation totals can be compared with crop yields.

4.3 Skewness in a distribution

A frequency distribution is said to be skew (or skewed) when its histogram or frequency curve is asymmetrical. If the skewness is *positive* the highest frequency (the mode) lies to the left of the centre of the distribution and the tail stretches out to the right towards the higher

(a)

(b)

(c)

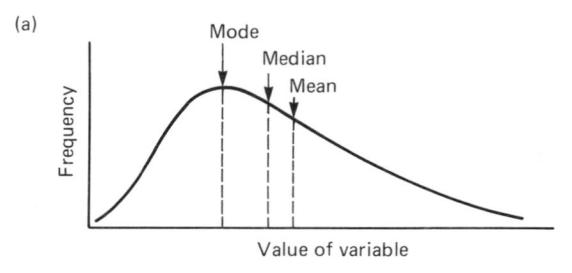

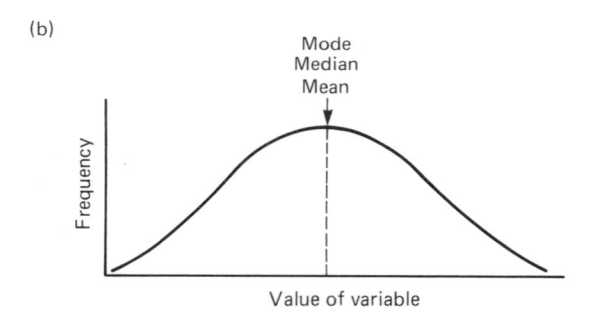

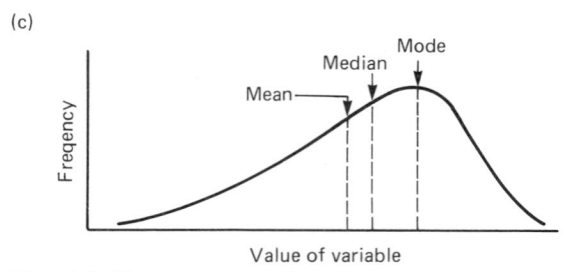

Fig. 4.5 Skewness in a distribution

values (Fig. 4.5(*a*)). When the skewness is *negative* the highest frequencies tend to be in the higher values of the variable and the tail extends to the left (Fig. 4.5(*c*).

The degree of skewness affects the relative location of the mean, the median and the mode on a graph. In Figure 4.5(*b*) the distribution is symmetrical. In this case the mean, the median and the mode are equal and all are situated at the centre of the distribution. In (*a*) the skewness is positive. The mode has a relatively low value and the mean has a higher value than the median. In (*c*) the opposite occurs. When the skew is negative the mean has a lower value than either the median or the mode.

The mean tends to be located on the same side of the mode as the longer tail of the distribution (the direction of skew). Thus a simple measure of skewness in a distribution is to subtract the value of the mode from the value of the mean: skewness = mean − mode. If there is no skew and the distribution is symmetrical (Fig. 4.5(*b*)) mean − mode = 0. If the mean lies to the right of the mode the value of the mean is greater than that of the mode (Fig. 4.5(*a*)), so mean − mode is positive and the skewness is also positive. If the mean lies to the left of the mode the value of the mode is greater than that of the mean (Fig. 4.5(*c*)) so mean − mode is negative and this indicates negative skewness. If the mean is equal to the mode, mean − mode = 0. There is no skewness; the distribution is symmetrical (Fig. 4.5(*b*)).

If desired, the skewness as calculated by mean − mode can be made comparable with that of other frequency distributions by dividing by the standard deviation, i.e. $\frac{\text{mean} - \text{mode}}{\text{standard deviation}}$. If preferred, it is possible to calculate the skewness of a distribution by subtracting the median from the mean, i.e. mean − median. As can be seen in Figure 4.5 this method is very similar to that outlined above.

4.4 The normal distribution

When histograms and frequency curves are constructed for continuous variables which are expressed in measurements, such as altitude, temperature, distance, weight, etc. they very often have a symmetrical, bell-like shape with high frequencies near the mean and lower frequencies to each side. Figure 4.6(a) is a histogram illustrating such a distribution of frequencies. Figure 4.6(b) is a frequency curve equivalent to Figure 4.6(a). This distribution is known as a *normal distribution*.

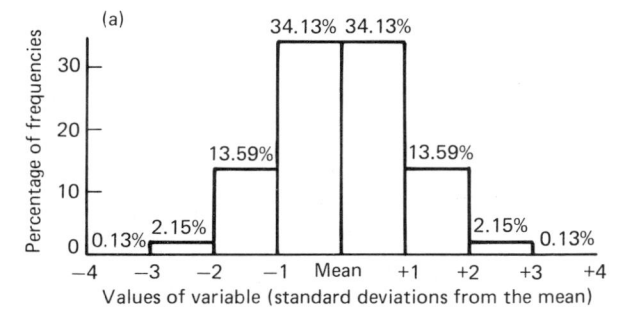

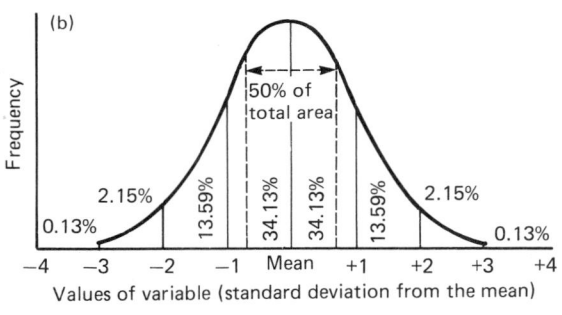

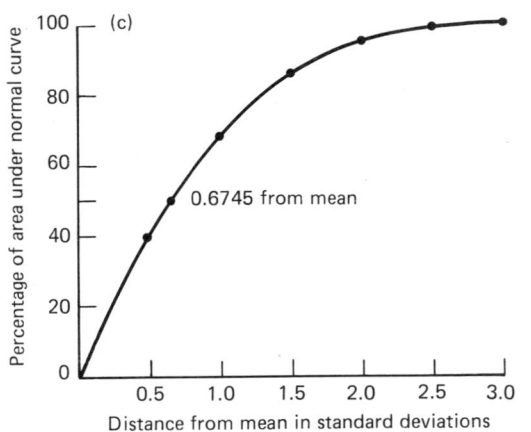

Fig. 4.6 The normal distribution

Normal distributions are symmetrical and unimodal. The arithmetic mean has a greater frequency than any other value. They cannot differ from one another in terms of their skewness or their kurtosis. A normal distribution is neither leptokurtic nor platykurtic.

A normal distribution is best described by the values of its arithmetic mean and its standard deviation. The mean locates the central value of the distribution which also has the peak frequency. The standard deviation describes the degree of dispersion of the distribution on each side of the mean. Just over two-thirds (68.27%) of the frequencies have a value between that of the mean and that of a deviation of 1 standard deviation (Fig. 4.6(c)). These frequencies are equally balanced on each side of the mean, about 34.13% occurring on each side. 95.45% of the frequencies lie within 2 standard deviations of the mean, with about 47.72 (i.e. 34.13 + 13.59) on each side of the mean. 99.73% of the frequencies lie within 3 standard deviations of the mean (Fig. 4.6(c)). It seems rather academic but 99.994% of the frequencies lie within 4 standard deviations of the mean (Fig. 4.6(b)). Actually the normal frequency curve extends *infinitely* in each direction and it can never reach the horizontal axis of Figure 4.6(b). Similarly the curve in Figure 4.6(c) can never quite reach a value of 100%. It is interesting to note that 50% of the total frequencies occur between the standard deviations -0.6745 and $+0.6745$ (Fig. 4.6(b) and (c)).

The area under the normal curve (Fig. 4.6(b)) represents the probability that an individual item selected at random will have a value in the range indicated by the horizontal axis of the curve. A randomly chosen value from a large normal distribution has about a 68% chance of having a value within 1 standard deviation of the mean, but it has less than a 5% chance of having a value that differs from the mean by more than 2 standard deviations. It has a 50% chance of being nearer to (or further away from) the mean than 0.6745 standard deviations (Fig. 4.6(b)).

Given the value of the mean and the standard deviation it is quite simple to deduce the major characteristics of a normal frequency distribution. Suppose a normally distributed set of temperature measurements has a mean value of 15°C and a standard deviation of 3°C. We can conclude that 68.27% of the temperature values lie between 1°C and 18°C. These differ from the mean (15°C) by less than 1 standard deviation. 13.59% of the measurements are likely to be between 12°C and 9°C and another 13.59% are likely to be between 18°C and 21°C. These differ from the mean by between 1 and 2 standard deviations. There is however a fairly remote possibility of there being a temperature that differs from the mean by more than 4 standard deviations, i.e. greater than 27°C or less than 3°C.

Exercises

1. Construct a histogram from each of the following sets of data:

    (a) Frequency   2 4 6 8
        Value of variable   1 2 3 4

    (b) Frequency   8 6 4 2
        Value of variable   1 2 3 4

    (c) Frequency   2 4 8 4 2
        Value of variable   1 2 3 4 5

    For each of these distributions calculate an index of skewness and comment on the meaning of the index in each case.

2. A normally distributed set of temperature measurements has a mean value of 10°C and a standard deviation of 3°C. Using Figure 4.6, give a more detailed description of the variability of these temperatures.

3. The two sets of figures given below represent the results of a map survey of the variations in the altitude of the ground surface in Area A and Area B. Fifteen locations were selected randomly in each area and the altitude in metres of each location was estimated. The altitude values for each area have been set out as an array, beginning with the lowest value.

Area A	Area B
120m	104m
135m	152m
162m	161m
170m	201m
190m	210m
195m	215m
204m	218m
208m	220m
212m	220m
220m	221m
220m	223m
232m	223m
240m	261m
252m	280m
260m	285m

    (a) For each set of altitude values, calculate the range and the quartile deviation. Comment on your results, noting any apparent anomaly.
    (b) Calculate the mean deviation of each set of altitude values and comment on the relationship in this example between the quartile deviation and the mean deviation.
    (c) Now calculate the standard deviation of each set of altitude values and comment on its relationship with the mean deviation.

4. The table below lists the population density per ha of 12 areas (A to L) of a county.

Area	Population density	Area	Population density
A	82	G	8
B	64	H	19
C	15	I	42
D	25	J	3
E	66	K	7
F	12	L	7

    To what extent does (a) the mean and (b) the median of the above density values provide a satisfactory summary of the population density of the whole county?

5. Construct a histogram from the data given below.

Class midpoint	1	2	3	4	5	6	7	8	9
Frequency of variable	2	4	8	6	4	3	2	2	1

    Using the histogram you have drawn, determine the value of:
    (a) The modal class and the mode;
    (b) The median class and the median;
    (c) The mean.
    Give a brief description of the histogram you have drawn.

5 Techniques Used in the Analysis of Spatial Point Patterns

Some of the techniques explained above for analysing numerical data can also be used to analyse two-dimensional spatial distributions. In such distributions dots can be placed on maps and their locations can be described, as on a graph, by quoting a value for the X-axis (e.g longitude on a map) and another value for the Y-axis (e.g. latitude on a map). For example, the location of the 'mean centre' of the point distribution in Figure 5.1 can be described by 'X = 5, Y = 4.9'.

Many different geographical features can be represented by dots on a map. The possibilities are limited by the scale of the map. With a large-scale map the location of shops, offices, churches and other features of similar scale can be indicated satisfactorily by dots. On an atlas map on the other hand a whole town or city may be represented by a single dot and its detailed features cannot be located. On some dot maps a dot may represent a number of items, such as a town with a population of 100 000 or even a million.

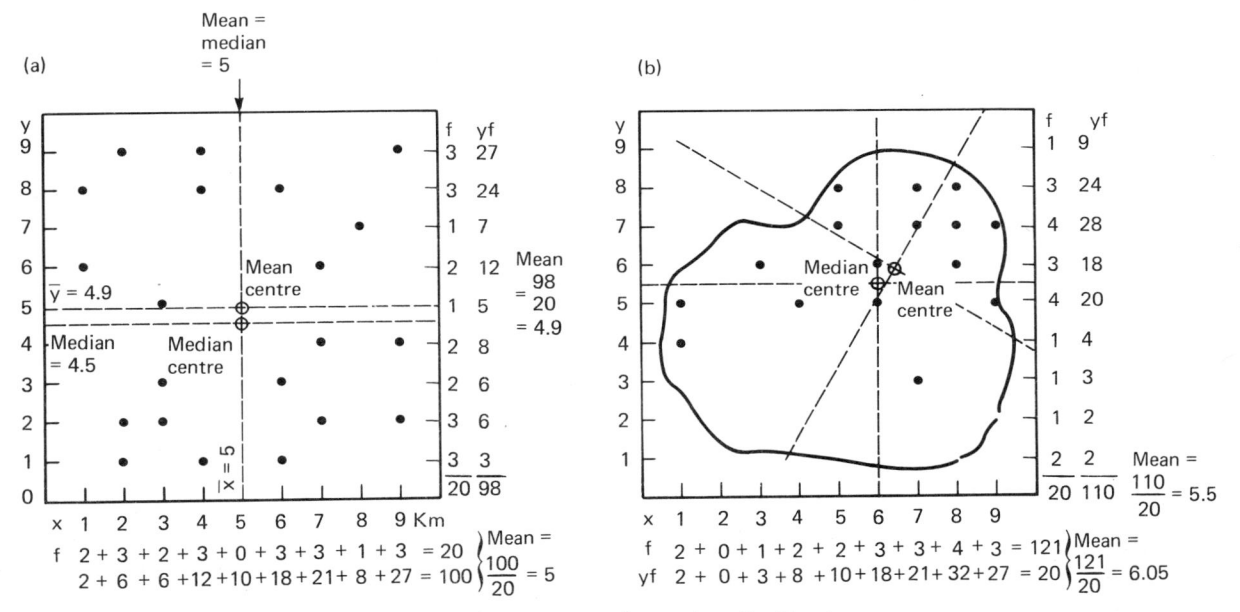

Fig. 5.1 The mean centre and the median centre of a point distribution

5.1 Averages

THE MEAN CENTRE AND MEDIAN CENTRE

Some of the techniques explained on pages 61 – 69 with reference to the analysis of numerical data can be applied to spatial data on dot maps. The measures of central tendency, the *mean centre* and the *median centre*, in a point distribution are the equivalent of the arithmetic mean and the median of a frequency distribution (pages 61 – 64). They indicate the approximate 'centre of gravity' of the distribution. In Figure 5.1(a) twenty dots are distributed in an area of 100 km² (10km × 10km). To find the mean centre of the distribution it is necessary to calculate the mean distance of the dots as measured along both the X-axis and the Y-axis. In Figure 5.1(a) the column headed f shows the frequency of dots at each distance along the Y-axis. There are three dots at a distance of 9 km from the origin and three more at a distance of 8 km. For each distance along the Y-axis the distance (y) has been multiplied by the frequency (f) to give the column headed yf. The sum of column yf divided by the number of dots gives the mean distance of the dots along the Y-axis. This mean distance is 4.9 km. A similar process has been applied to the X-axis and the mean distance of the dots along the X-axis is 5 km. Two lines representing these two mean distances have been drawn on the diagram. Where these lines cross is the mean centre of the dot distribution.

The *median centre* is the $\frac{n+1}{2}$ th dot along both the X-axis and the Y-axis. Since there are 20 dots the median is the 10.5th dot along each axis. Since there are no decimals of a dot this means that there will be 10 dots on each side of the median line parallel to the Y-axis and also 10 dots on each side of that parallel to the X-axis. Where these two lines cross is the median centre. Its position can change slightly if the two median lines are rotated slightly.

Figure 5.1(b) shows a more realistic example because an irregularly shaped area is indicated that could be an island, a county, a country, an upland area, an area of farmland, a central business district or any other area of any shape or scale. The grid superimposed on this area is large enough to enclose it completely. The mean distances along the X and Y axes have been calculated exactly as before, and the mean centre of the dot distribution has been indicated. The median lines however have been rotated so that there are equal numbers of dots to the north-east and the south-west and also to the north-west and south-east. The median centre itself sometimes shifts its position slightly if the orientation of the median lines is changed. The mean centre however can be determined using construction lines at an angle, provided of course that they are perpendicular to each other. The dot distribution in Figure 5.1(b) is skewed. There is a much greater concentration of dots in the north-east than in the south-west, so the direction of skew is towards the south-west. Figure 4.5 (page 68) shows that in a skewed frequency distribution the median is located nearer than the mean to the values with the higher frequencies. In a skewed spatial distribution the median centre lies nearer than the mean centre to the denser concentration of points.

When dealing with a point distribution so large that it would be laborious to consider each point individually it is better to group the data into classes before beginning to process it. In Figure 5.2 there are 50 points but it would be possible to handle a much larger number with very little increase in difficulty. The point pattern is treated as a grouped frequency distribution on both the X axis and the Y axis. Along the axes the dots in the columns (X axis) and the rows (Y axis) are all assumed to have the value of the mid-point of the column or the row. Thus the 8 dots in the first column on the X axis are each given a value of 1 and the 9 dots in the second column are each given a value of 3. These values and frequencies are summarised by the histogram beneath the dot distribution. To find the mean distances of the dots along the X axis the distance of the mid-point of each column (x) is multiplied by the frequency (f) to give a value (xf) for each column. In Figure 5.2 these values are successively 8 (1 × 8), 27 (3 × 9), 65 (5 × 13), 105 (7 × 15) and 45 (9 × 5). The sum of the values of xf is therefore 250. This is divided by the number of points (50) and this gives a mean distance of 5 km for the dots along the X axis. The same process is carried out along the Y axis and the mean distance here is 4.4 km. The mean centre of the distribution therefore is the point that is located at 5 km along the X axis and 4.4 km along the Y axis.

It is possible to find the median centre by the simple method explained above but the simple process can prove quite difficult to apply when there is a large number of dots crowded closely together.

TECHNIQUES USED IN THE ANALYSIS OF SPATIAL POINT PATTERNS

The mode and modal class

Another description of central tendency which can be used in relation to the distribution in Figure 5.2 is the *mode*. Since only one dot can occupy at a particular location there can be no precise 'mode', but it is possible to identify the *modal class*. This is the square that contains the greatest number of dots. In Figure 5.2 the modal class has a value of 6 and it is located 7 km along the X axis and 5 km along the Y axis. It also coincides with the modal classes of the two histograms.

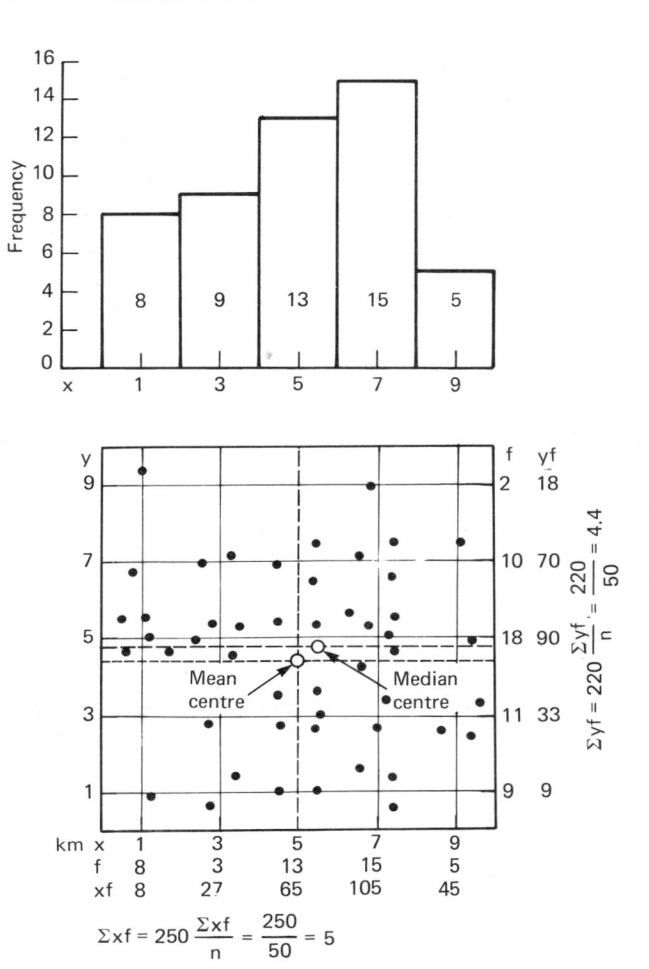

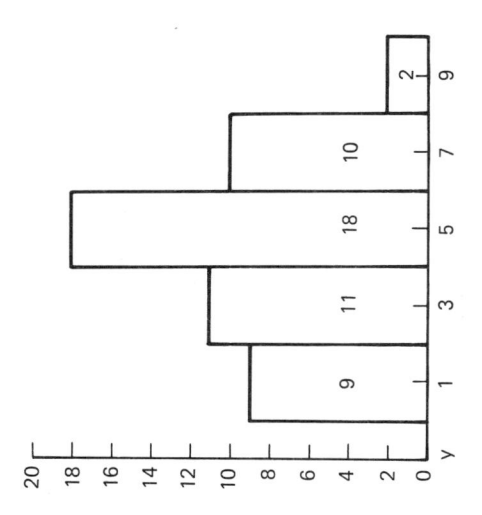

Fig. 5.2 The mean centre and median centre of a large dot distribution

5.2 Measures of spatial dispersion

The standard distance

It would be possible to adapt either the mean deviation or the standard deviation to describe the degree of dispersion in a point distribution. In practice it is the standard deviation that is usually used. In a point distribution the term 'dispersion' usually means the extent to which the various point locations deviate from the position of the mean centre just as 'dispersion' in a frequency distribution is the extent to which values deviate from the mean value. In a point distribution dispersion is measured by the *standard distance* which is the spatial equivalent of the standard deviation. It is the square root of the mean value of the squared distances between the mean centre and the various dots in the distribution—in short: $\sqrt{\frac{\Sigma d^2}{n}}$.

Figure 5.3(a) shows a spatial distribution of 20 points. Figure 5.3(b) shows the location of the mean centre in relation to the point distribution. Figure 5.3(c) shows the calculations that establish the value of the standard distance. First of all the position of the mean centre of the distribution is determined, as explained above. Then the distance from each point to the mean centre is carefully measured. This gives the column headed d in (c). Each of these values is then squared to give the column headed d^2. The values in the d^2 column are then summed, and the total value in this case is 520.8 km. This total is now divided by 20 (the number of points on the map) and this gives 26.04. Finally the square root of 26.04 (i.e. 5.1 km) is the standard distance. A circle with this radius has been drawn on (b). Figure 5.4 shows the result of applying the same procedure to a random distribution of 20 points generated by the use of random numbers. Although the study areas are the same size and shape and have equal numbers of dots, the random distribution clearly differs greatly from that of Figure 5.3. The standard distance is 6.54 km compared with 5.1 km. There are only 10 points (50% of the total) within the standard distance of the mean centre instead of 12 (60% of the total). Although the overall density of points is the same in both distributions the density of points within the 1 standard distance circle in Figure 5.3 is twice that in Figure 5.4. It is clear that there is a greater concentration of points near the mean centre in Figure 5.3 than in Figure 5.4.

This is not really very surprising since a random point distribution is created by using random numbers each of which has an equal chance of being placed at any location within the study area. However the point distribution in Figure 5.3 cannot be described as the equivalent of a normal frequency distribution. In a 'normal' point distribution one would expect about 68% of the points to lie within one standard distance of the mean centre. This would mean that 13 or 14 dots would occur within the circle. In fact there are only 12. Therefore the distribution in Figure 5.3 is rather more dispersed than a normal distribution.

Chi-square

Many point distributions show little or no tendency to cluster around the mean centre, as in Figure 5.4. Others may be regularly spaced, for example in a distribution of street junctions in a gridiron street pattern. Some may be randomly spaced. Others may consist of groups of points clustered here and there as in the

Calculation of standard distances

Point	Distance from mean centre (km)	
	d	d^2
A	7.4	54.8
B	7.5	56.3
C	4.9	24.0
D	4.3	18.5
E	6.5	42.3
F	4.1	16.8
G	2.2	4.8
H	5.9	34.8
J	2.5	6.3
K	1.1	1.2
L	1.7	2.9
M	6.2	38.4
N	3.4	11.6
O	1.4	2.0
P	4.5	20.3
Q	4.5	20.3
R	3.5	12.3
S	7.5	56.3
T	6.6	43.6
U	7.3	53.2
		520.8

$$\frac{520.8}{20} = 26.04$$

$$\sqrt{26.04} = 5.1$$

Standard distance = 5.1 km

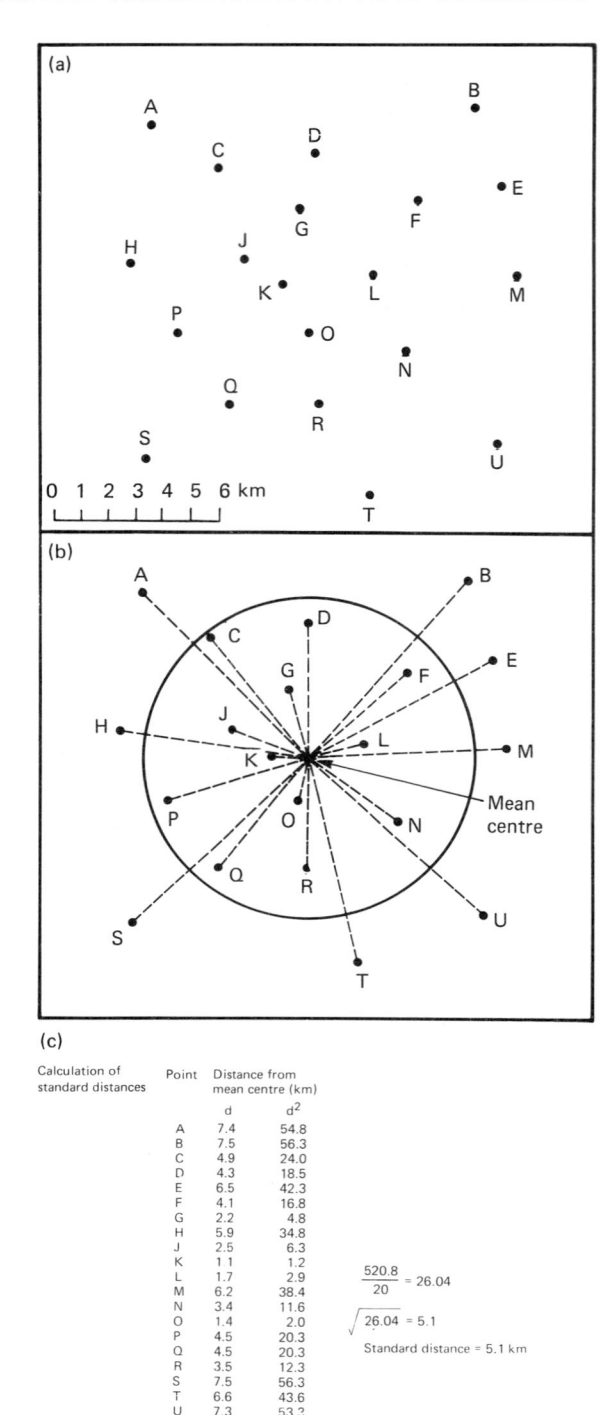

Fig. 5.3 The calculation of the standard distance of a point distribution

TECHNIQUES USED IN THE ANALYSIS OF SPATIAL POINT PATTERNS

case of the distribution of houses in sparsely populated rural areas. In such cases the concept of a 'normal' point distribution has very little relevance. It is possible however to analyse these types of dot distributions by using the *chi-square* technique. Normally this technique is used to compare the actual distribution of points with a random distribution of the same number of points. First of all a *null hypothesis* (H_0) is formulated to the effect that there is no significant difference between the *observed* pattern of points and the *expected* pattern which is usually regarded as a random pattern. A random spatial point pattern is fairly irregular in detail, with a number of small clusters, but it tends to be regular when considered at a larger scale. This is because, in a random distribution, each location in the study area has an equal chance of being selected by a random sampling number. Consequently equally sized compartments in a study area will tend to have equal numbers of points, as in the case of a regular distribution, but the random distribution will be more variable in detail.

The aim of using the chi-square test is to find out whether the observed frequencies agree with or differ from the theoretical (expected) frequencies.

Figure 5.5 shows the method of calculating the value of chi-square. The letters, A to H in the table, refer to the areas, A to H, above the table. In the column headed O are listed the numbers of points in each of areas A to H on the map (i.e. the observed frequencies). The total number of points in this case is 40. Column E contains the list of expected frequencies in each of the areas A to H assuming that the points are randomly spaced. In the column headed O − E each of the expected frequencies (E) is subtracted from the corresponding observed frequency, and in the last column each value of O − E is squared. The relevant values are then inserted into the expression for chi-square, i.e. $\sum \frac{(O-E)^2}{E}$. The resultant value for chi-square is 2.4. The tables of Figure 5.5(*b*) and (*c*) have been processed in a similar way and their chi-square values are 8.0 and 32.4 respectively. The point distribution in Figure 5.5(*a*) clearly matches the expected distribution quite closely. No value in the (O − E)² column is greater than 4 and in two cases it is zero.

(a) [map of points A–U with scale 0–6 Km]

(b) [map showing points around a mean centre within a circle]

(c) Calculation of standard distance

Point	Distance from mean centre (km) d	d^2
A	9.4	88.4
B	8.0	64.0
C	8.8	77.4
D	6.5	42.3
E	8.9	79.2
F	7.1	50.4
G	2.8	7.8
H	3.2	10.2
J	7.9	62.4
K	2.6	6.8
L	4.7	22.1
M	1.3	1.7
N	3.7	13.7
O	5.5	30.3
P	4.0	16.0
Q	6.3	39.7
R	9.1	82.8
S	7.5	56.3
T	7.2	50.4
U	7.3	53.3
		855.2

$\frac{855.2}{20} = 42.76$

$\sqrt{42.76} = 6.54$

Standard distance = 6.54

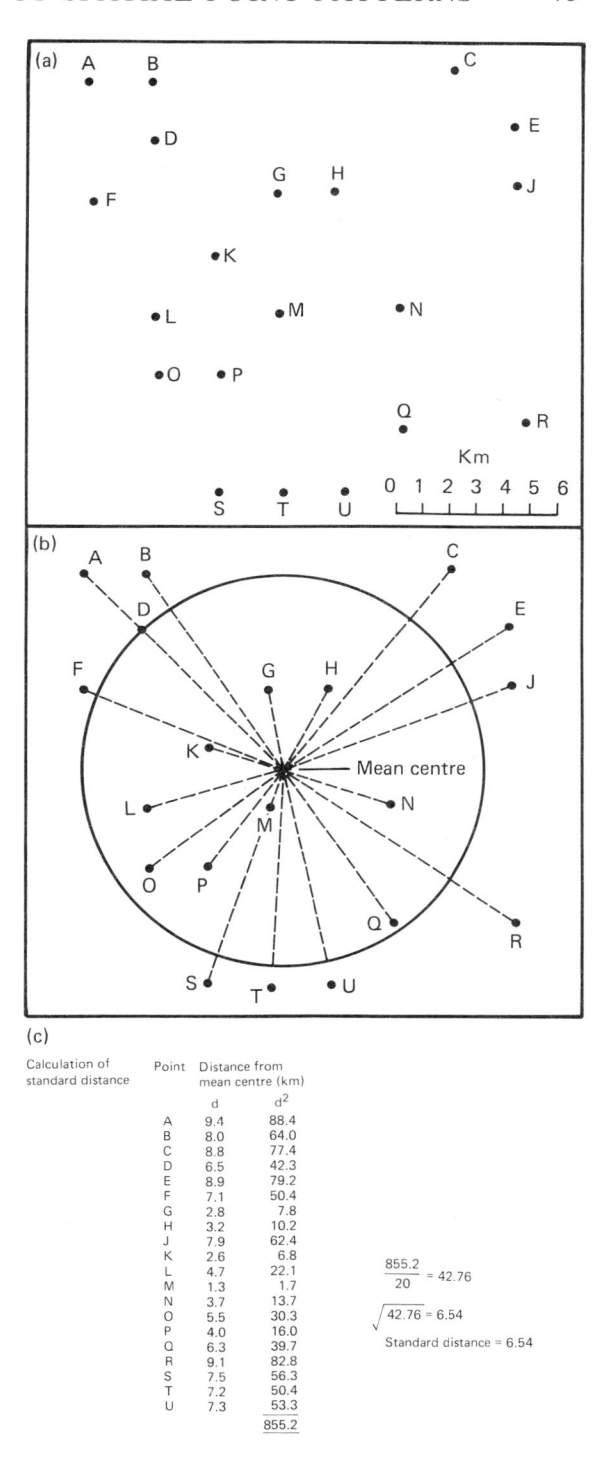

Fig. 5.4 The standard distance of a random point distribution

The chi-square value is very low indeed at 2.4. On the other hand the distribution shown in Figure 5.5(c) is very different from the expected distribution and has easily the highest value of chi-square (32.4). In this case there appears to be a greater tendency to reject the null hypothesis since there is no significant difference between the observed pattern of dots and the expected pattern. The strength of the relationship between the expected distribution

(a)

	O	E	O − E	$(O-E)^2$
A	4	5	−1	1
B	6	5	+1	1
C	4	5	−1	1
D	5	5	0	0
E	3	5	−2	4
F	5	5	0	0
G	7	5	+2	4
H	6	5	+1	1
	40	40	0	12

$$\chi^2 = \Sigma \frac{(O-E)^2}{E} = \Sigma \frac{12}{5} = 2.4$$

(b)

	O	E	O − E	$(O-E)^2$
A	8	5	+3	9
B	6	5	+1	1
C	3	5	−2	4
D	7	5	+2	4
E	1	5	−4	16
F	4	5	−1	1
G	7	5	+2	4
H	4	5	−1	1
	40	40	0	40

$$\chi^2 = \Sigma \frac{(O-E)^2}{E} = \Sigma \frac{40}{5} = 8.0$$

(c)

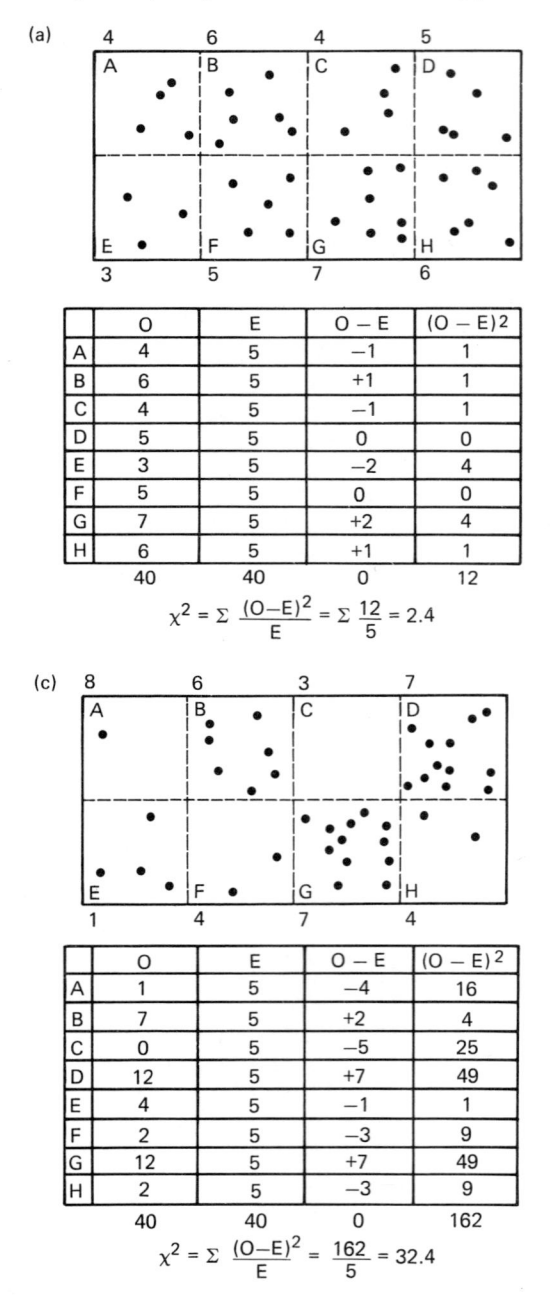

	O	E	O − E	$(O-E)^2$
A	1	5	−4	16
B	7	5	+2	4
C	0	5	−5	25
D	12	5	+7	49
E	4	5	−1	1
F	2	5	−3	9
G	12	5	+7	49
H	2	5	−3	9
	40	40	0	162

$$\chi^2 = \Sigma \frac{(O-E)^2}{E} = \frac{162}{5} = 32.4$$

Fig. 5.5 The calculation of chi-square

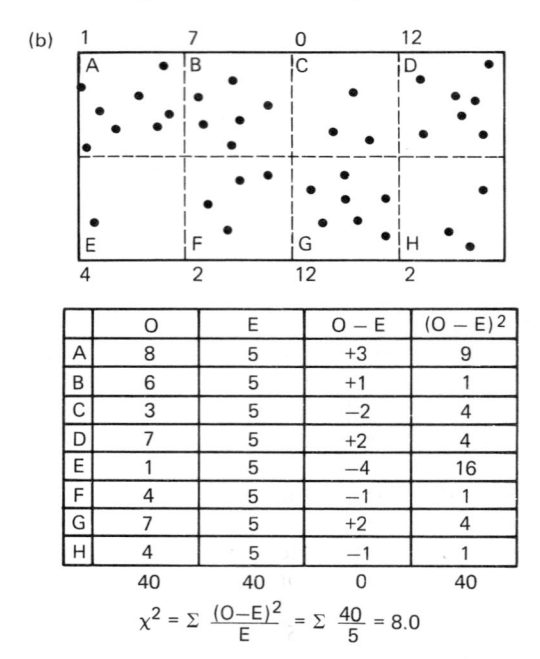

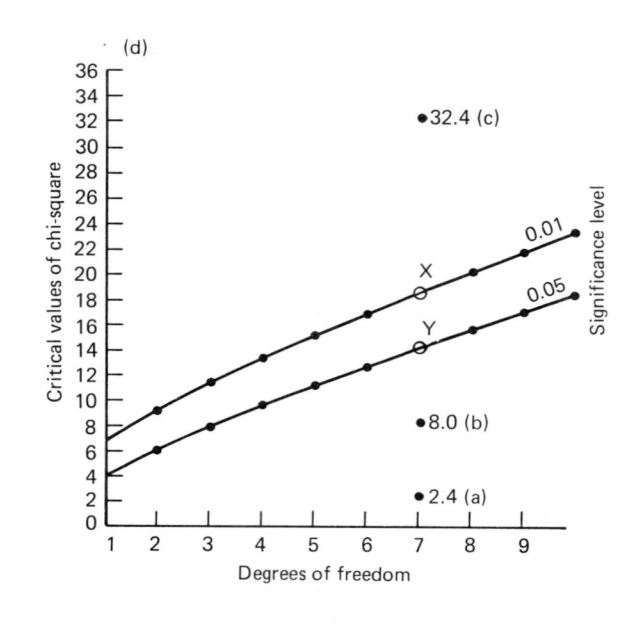

and the observed distribution can be tested by using the graph (Fig. 5.5(d)). This graph links the chi-square value with the degrees of freedom and the significance level. In this simple form of the chi-square test the number of degrees of freedom is one less than the number of frequencies considered (i.e. n − 1). In this particular case it is 7 (i.e. 8 − 1). The significance level indicates the probability that the difference between 0 (observed frequency) and E (expected frequency) could arise by chance. Probabilities of 0.01 (1 in 100) or 0.05 (5 in 100) are commonly used. On the graph the *critical value* of chi-square is located where the line showing the level of significance crosses the line showng the degrees of freedom relevant to the data. In all three of our distributions, using the 0.01 (1 in 100) significance level the critical value for chi-square is slightly over 18 (at x). Using the 0.05 significance level the critical value would have been almost exactly 14 (at Y).

The null hypothesis that there is no significant difference between the observed pattern of points and the expected random pattern is rejected if the value of chi-square is greater than this critical value. It is therefore rejected in the case of Figure 5.5(c) whose chi-square value is 32.4, much greater than either of the critical values of chi-square. But the null hypothesis cannot be rejected in the other two cases whose chi-square values are well below the critical values. So Figures 5.5(a) and 5.5(b) may be classified as similar to a random (or regular) distribution.

In using this technique it is important that the grid squares used are of equal size and scale. A coarser grid with large divisions can reduce the value of chi-square. One important advantage of the chi-square technique is that it allows very large distributions of dots to be handled fairly easily. Their characteristics can be summarisd by the chi-square index and distributions of points can be compared with one another.

Nearest-neighbour analysis

In point distributions it is also possible to analyse the detailed relationships between individual points in the pattern. In *nearest-neighbour analysis* we are concerned with the varying degrees of clustering, regularity or randomness in the relative locations of the points which make up the pattern. These qualities are often judged by guesswork in the course of analysing maps, but nearest-neighbour analysis is a means of standardising the assessment and making it objective. In nearest-neighbour analysis a particular point pattern is compared with a random point distribution which has the same number of points in the same size of area. The random point pattern is referred to as the '*expected distribution*' ($\bar{r}E$). The point pattern under study is referred to as the '*observed distribution*'. This is abbreviated to $\bar{D}$ obs.

First of all it is necessary to calculate the average distance between the points and their nearest neighbours in a random pattern of the same size (i.e. $\bar{r}E$). The simplest way to do this is to use the relationship:

$$\bar{r}E = 0.5 \sqrt{\frac{\text{Size of study area}}{\text{Number of points}}}$$

This calculation is shown in Figure 5.6(a). The size of the study area is 100 km^2 and the number of points is 10. This gives an $\bar{r}E$ value of 1.58 km which is the average nearest-neighbour distance in a random distribution of 10 points located in an area of 100 km^2.

Next it is necessary to measure on the map the distances between each of the points and its nearest-neighbour (also in kilometres). These distances are listed in the tables in Figure 5.6(a). The average of the *observed* nearest-neighbour distances ($\bar{D}$ obs) is found to be 2.87 km. This is calculated by dividing the total of the nearest-neighbour distances (i.e. 28.7 km) by the number of points (i.e. 10). We have now calculated the expected average nearest-neighbour distance ($\bar{r}E$) in a random distribution of this size (i.e. 1.58 km) and also the observed nearest-neighbour distance ($\bar{D}$ obs) from the map (i.e. 2.87 km). The nearest-neighbour statistic (Rn) is now calculated by dividing $\bar{D}$obs by $\bar{r}E$. This gives a value of 1.82 (Fig. 5.6(a)). In (a) there is a scale showing the significance of the Rn values. If the Rn value is 1.0 the point distribution is similar to a random one. This however should not be interpreted to mean that the distribution has been created by influences that have operated in a random way. We are measuring only descriptive aspects of the distributions. However, since an Rn value of 1.0

(a)

Location	Nearest neighbour	Nearest neighbour (distance km)
A	E	3.1
B	D	2.6
C	B	2.8
D	B	2.6
E	F	2.7
F	E	2.7
G	D	2.7
H	E	3.0
J	F	3.5
K	G	3.0

Total = 28.7
$\bar{D}$ obs = 2.87 km

Expected average nearest neighbour distance ($\bar{r}E$)

$= 0.5 \sqrt{\dfrac{\text{Area}}{\text{Number of points}}}$

$= 0.5 \sqrt{\dfrac{100}{10}}$

$= 0.5 \sqrt{10}$

$= 0.5 \times 3.16$

$\bar{r}E = 1.58$ km

$Rn = \dfrac{2.87}{1.58} = 1.82$

(b)(i)

$\bar{D}$ obs = 1 km
Expected average nearest neighbour distance

$= 0.5 \sqrt{\dfrac{\text{Area}}{\text{Number of points}}}$

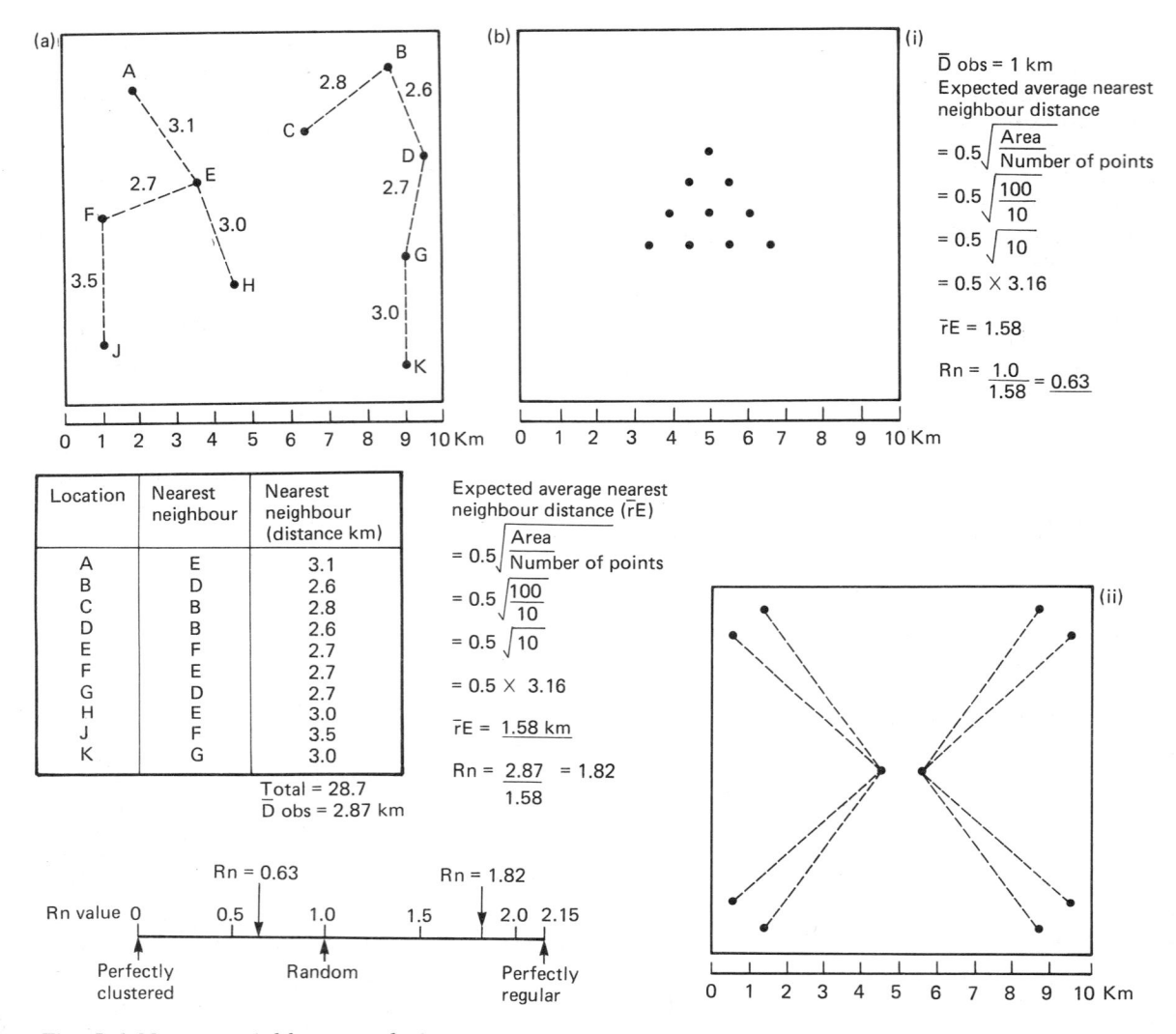

$= 0.5 \sqrt{\dfrac{100}{10}}$

$= 0.5 \sqrt{10}$

$= 0.5 \times 3.16$

$\bar{r}E = 1.58$

$Rn = \dfrac{1.0}{1.58} = \underline{0.63}$

Rn value scale:
- 0 — Perfectly clustered
- 1.0 — Random
- 2.15 — Perfectly regular
- Rn = 0.63 (marked at ~0.63)
- Rn = 1.82 (marked at ~1.82)

Fig. 5.6 Nearest-neighbour analysis

means that the observed nearest-neighbour distance is equal to the expected distance the distribution resembles a random one. If the Rn value is greater than 1.0, as in this case, the distribution is more regular than a random one. The maximum value for Rn is 2.15 which is the value for a perfectly regular point distribution with points at the apices of equilateral triangles covering the whole area. When one looks at the map in Figure 5.6(a) it seems correct to describe the point distribution as 'regular'. Points are fairly evenly distributed over much of the area.

Problems can arise when using nearest-neighbour analysis to describe a clustered point distribution. In Figure 5.6 (b)(i) all the points have a nearest-neighbour distance of 1 km, so their average nearest-neighbour distance ($\bar{D}$ obs) must be 1 km. The value of the expected nearest-neighbour distance ($\bar{r}E$) is 1.58, exactly the same as in the distribution of Figure 5.6(a) because the total area and the number of points are both the same. The value of Rn in (b)(i) is only 0.63. This indicates a distribution considerably more clustered than a random one. Figure 5.6(b)(i) would certainly be described as a strongly clustered point distribution, all ten points being concentrated in the centre of the area. Figure 5.6(b)(ii) however has exactly the same Rn value as (b)(i). All its points have a nearest-neighbour distance of 1 km. This distribution consists of five sets of *reflexive pairs*. A reflexive pair is a set of two points each of which is the other's nearest neighbour. Reflexive pairs are particularly numerous in random-like distributions.

To distinguish between (b)(i) and (b)(ii) we should have to take into account second-nearest neighbours (marked by dashed lines). The second-nearest neighbours in (b)(ii) are at a much greater distance than those in (b)(i), so we can conclude that the distribution in (b)(ii) is less clustered than that in (b)(i).

Exercises

1. Refer to Figure 5.2 which is a fairly large point distribution. The diagram below shows a similar type of distribution.

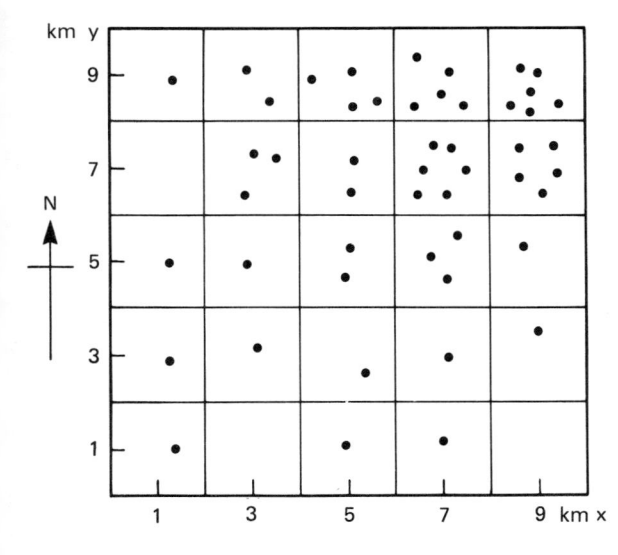

   (a) Make a copy of the 25 square compartments and write in each compartment the number of dots that it contains.
   (b) Draw two histograms, as in Figure 5.2, and identify and label the modal class.
   (c) Mark the location of the mean centre and the median centre of this distribution, and identify the direction of skew.

2. (a) On graph paper make a copy of the point distribution below which is composed of 20 dots. Plot on your map the mean centre and the median centre of this distribution.
   (b) To what extent is this point distribution skewed?

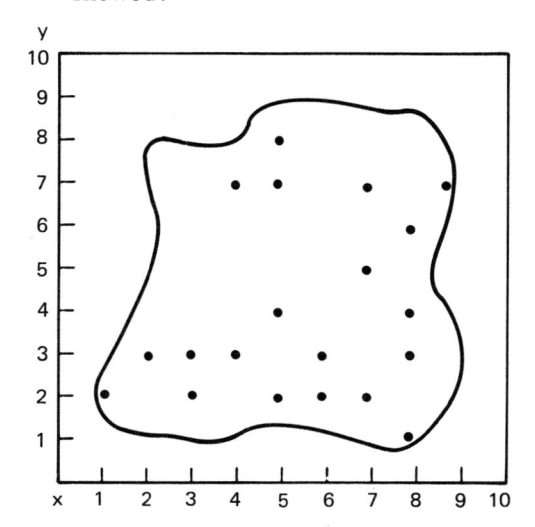

3. The map shows the distribution of road junctions in a rural area. The density of road junctions may be regarded as approximately proportional to the density of the road network. The approximate mean centre of the point distribution is indicated by X. Calculate:

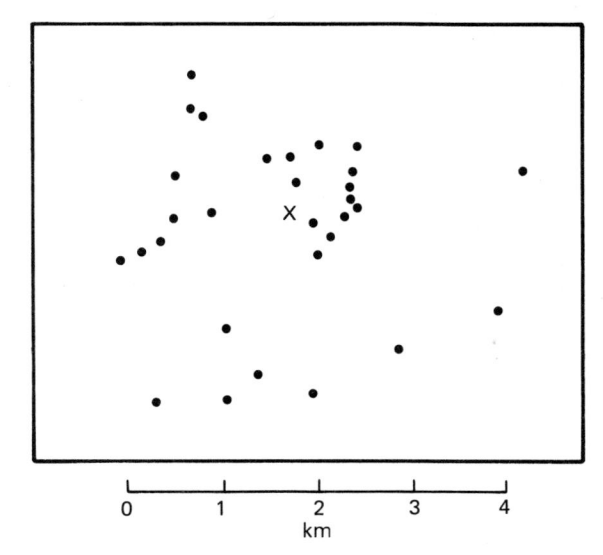

(a) The mean distance of the road junctions from X;
(b) The standard distance of the road junctions from X.
4. The map below shows the distribution of villages in an area of chalk downland. Referring to Figure 5.6, describe the overall characteristics of this settlement pattern.

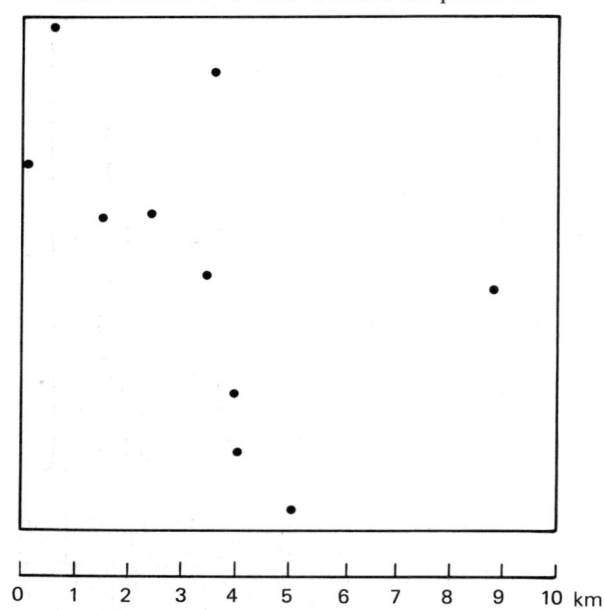

5. Using the chi-square technique, test the null hypothesis that there is no significant difference between the point distribution shown in the diagram below and a random distribution.

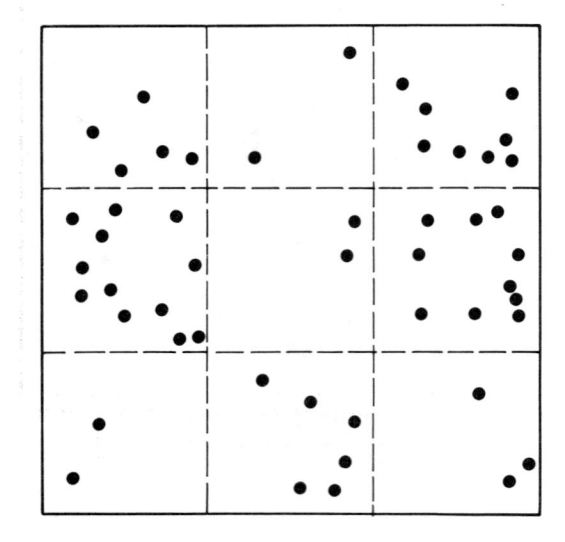

6 Techniques Used in the Analysis of Networks

Lines and points can be combined to create networks. Two important types of network in geography are the stream networks that exist in drainage basins and transport networks made up of interlacing patterns of roads, railways and air and sea routes.

6.1 Drainage networks

In a drainage basin a network of stream channels conveys much of the precipitation that falls in the area bounded by the watershed to the basin outlet. Near the watershed it may be quite difficult to identify the point at which a stream begins (its source) since this point can vary as the relationship between precipitation and evapotranspiration changes. If a stream is considered to form part of the drainage network it should flow permanently and should be connected to the rest of the network. Various techniques have been developed to describe and compare the characteristics of drainage basins and their stream networks. Some of these refer to the overall characteristics of the drainage basin. Others refer to the more detailed characteristics of the streams within the drainage basin.

Some of the ways of describing the characteristics of a drainage network as a whole are illustrated in Figure 6.1 in respect of two contrasting networks.

The form factor

The *form factor* compares the area of the basin (basin area) with the area of a square whose sides are equal in length to the distance from the basin outlet to the most distant point on the basin's perimeter (i.e. basin length^2). In other words the basin is being compared with a square each of whose sides is the same length as the maximum basin length. Network A has an area that is 61% (0.61) of the area of the square. The area of Network B on the other hand is only 33% (0.33) of the area of the square. Network A therefore has a more compact shape than Network B. The maximum value for the form factor is 1.0 which means of course that the drainage basin would be square (Fig. 6.1(*a*)). The *shape index* is an alternative way of comparing the two shapes which gives very similar results (Fig. 6.1(*a*)).

Basin relief

Another measure that is useful when comparing drainage basins is the *basin relief* (Fig. 6.1(*b*)). This is simply the difference in altitude between the highest point in the basin and the lowest point. The lowest point will be at the outlet of the basin. Basin relief can have a strong influence on the flow directions and the detailed channel patterns of the streams of the basin. For example, steep slopes will tend to result in straight streams.

Drainage density

Finally the *drainage density* of the basin can be measured. This is the length of drainage channel per unit area of the basin. First of all the total length of all the streams in the basin is measured. Then this is divided by the area of the basin (Fig. 6.1(*b*)). Network A has a stream length of 37.5 km and Network B's is only 27 km, but because Network B (basin area 33 km^2) is so much smaller than Network A (basin area 61 km^2), Network B has the greater drainage density (0.8 km/km^2 compared with 0.6 km/km^2).

STREAM CHANNELS

Techniques are also available for the description and comparison of more detailed characteristics of drainage networks. These concentrate atten-

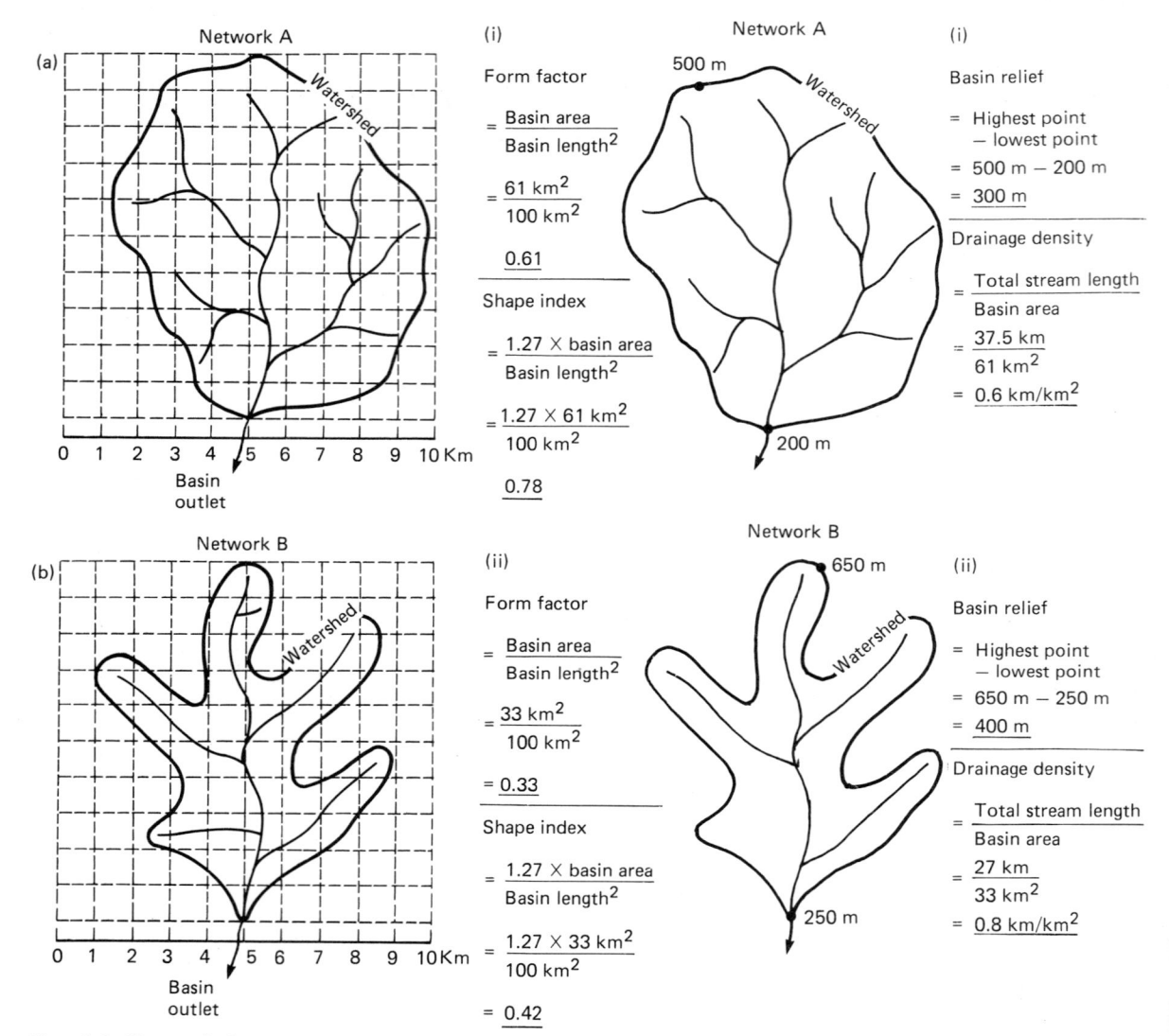

Fig. 6.1 General characteristics of a drainage basin

(a) Network A

Form factor
$$= \frac{\text{Basin area}}{\text{Basin length}^2}$$
$$= \frac{61 \text{ km}^2}{100 \text{ km}^2}$$
$$= \underline{0.61}$$

Shape index
$$= \frac{1.27 \times \text{basin area}}{\text{Basin length}^2}$$
$$= \frac{1.27 \times 61 \text{ km}^2}{100 \text{ km}^2}$$
$$= \underline{0.78}$$

Network A

Basin relief
$$= \text{Highest point} - \text{lowest point}$$
$$= 500 \text{ m} - 200 \text{ m}$$
$$= \underline{300 \text{ m}}$$

Drainage density
$$= \frac{\text{Total stream length}}{\text{Basin area}}$$
$$= \frac{37.5 \text{ km}}{61 \text{ km}^2}$$
$$= \underline{0.6 \text{ km/km}^2}$$

(b) Network B

Form factor
$$= \frac{\text{Basin area}}{\text{Basin length}^2}$$
$$= \frac{33 \text{ km}^2}{100 \text{ km}^2}$$
$$= \underline{0.33}$$

Shape index
$$= \frac{1.27 \times \text{basin area}}{\text{Basin length}^2}$$
$$= \frac{1.27 \times 33 \text{ km}^2}{100 \text{ km}^2}$$
$$= \underline{0.42}$$

Network B

Basin relief
$$= \text{Highest point} - \text{lowest point}$$
$$= 650 \text{ m} - 250 \text{ m}$$
$$= \underline{400 \text{ m}}$$

Drainage density
$$= \frac{\text{Total stream length}}{\text{Basin area}}$$
$$= \frac{27 \text{ km}}{33 \text{ km}^2}$$
$$= \underline{0.8 \text{ km/km}^2}$$

tion on the stream channels rather than the drainage basin as a whole. At a fairly superficial level drainage networks can be described by adjectives such as 'dendritic' (tree-like), centripetal, trellis and others, but this kind of classification depends mainly on subjective judgment.

Stream order

In order to classify drainage channels objectively the stream segments can be divided into a hierarchy of *stream orders*. There are several ways of doing this but the most popular method is that devised by Strahler. In Strahler's system the 'fingertip' channels with no tributaries are referred to as first-order streams. These are labelled 1 in Networks A and B (Fig. 6.2). When two first-order streams meet they create a second-order stream which can receive first-order tributaries without changing its order. When two second-order streams meet they create a third-order stream which can receive first and second-order streams without changing its order. This process continues into the higher orders. The confluence of two streams of the

(a) Network A   Network B

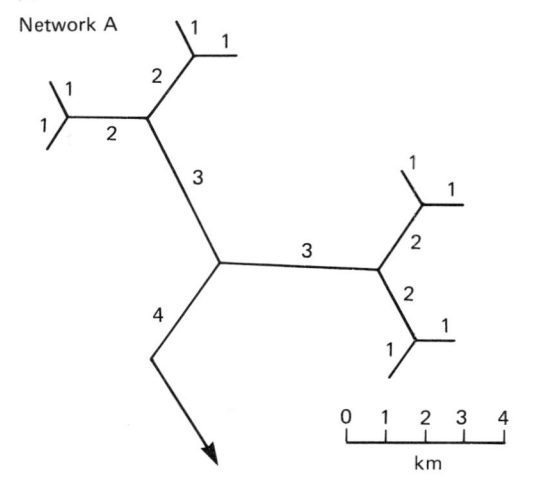

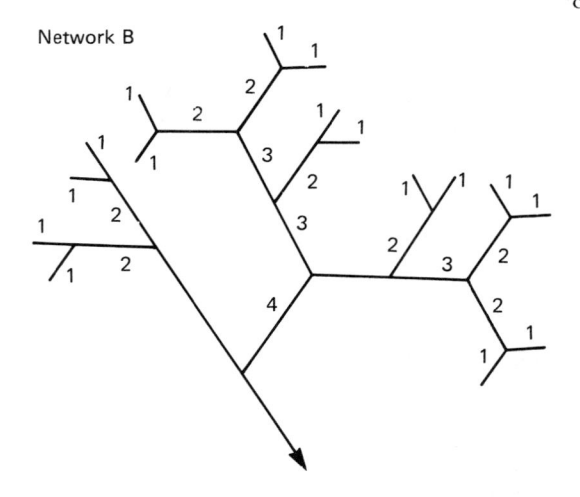

Network A

Stream order	Number of segments	Bifurcation ratio
1	8	
		2.0
2	4	
		2.0
3	2	
		2.0
4	1	

Mean segment length (km)	Length ratio
1	
	2.0
2	
	2.0
4	
	1.5
6	

Network B

Stream order	Number of segments	Bifurcation ratio
1	16	
		2.0
2	8	
		2.7
3	3	
		3.0
4	1	

Mean segment length (km)	Length ratio
1	
	2.0
2	
	2.0
4	
	1.5
6	

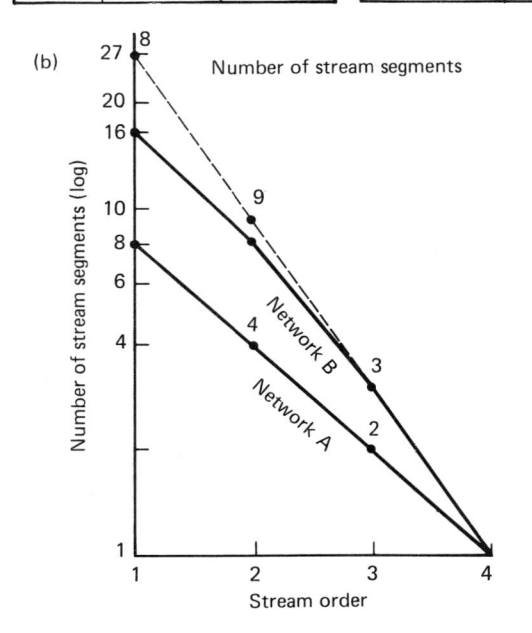

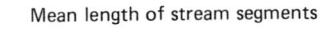

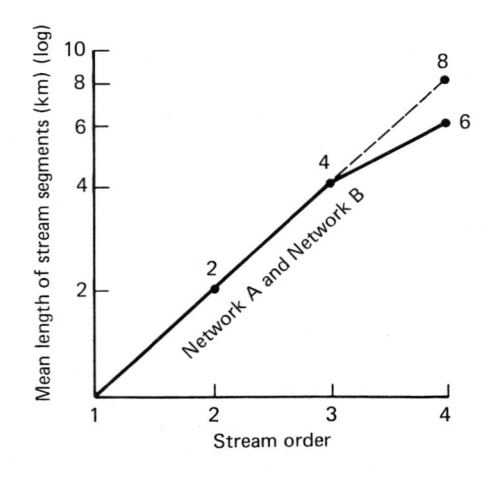

Fig. 6.2 Detailed characteristics of a drainage basin

same order creates a stream segment of the next higher order. The system can be summarised as: 1 + 1 = 2, 2 + 2 = 3, 3 + 3 = 4, etc.

In Network A (fig. 6.2(*a*)) there are eight stream segments of the first order, with no tributaries. Each pair of these first-order streams creates a second-order stream, so there are four second-order streams. These, in turn, create two third-order streams which, in turn, create one fourth-order stream. This relationship is summarised in the table referring to Network A. Network A has the absolute minimum number of streams in each order. Network B has many more streams but its highest order stream is only of the fourth-order, just the same as in Network A. Network B has a number of 'redundant' streams that do not affect the stream ordering.

The bifurcation ratio

When the stream segments have been classified into orders it is possible to calculate the *bifurcation ratio*. This is the ratio of the number of streams in any order to the number in the next higher order. In Network A the bifurcation ratio is always 2. The lower order always has twice as many stream segments as the next higher order. In Network B, on the other hand, the bifurcation index varies. There are three times as many stream segments in the third-order as there are in the fourth-order, but in the lower orders the index is lower.

In the semi-logarithmic graph in Figure 6.2(*b*) a constant bifurcation ratio is indicated by a straight line as in the case of Network A. This indicates a *constant rate* of increase in the number of stream segments with decreasing stream orders. Also, in Figure 6.2(*b*) it can be seen that Network B has relatively too few stream segments in the lower stream orders to maintain a constant bifurcation ratio.

It is also possible to study drainage networks from the point of view of the relative length of the stream segments belonging to the various stream orders. First of all the average length of the stream segments of the various stream orders is calculated. It is then possible to calculate the *length ratio*. This is similar to the bifurcation ratio in that it is the ratio of the average length of the streams in any order to the average length in the next order. For example, the average length of second-order streams in Network A is 2 km and the average length of first-order streams is 1 km. So the length ratio between second and first order streams is 2.0. It should be noted that the mean stream segment length tends to be greater in high order streams, unlike the number of stream segments. Values of the length ratio for the various pairs of stream orders can be plotted on a semi-logarithmic graph which indicates the rate of change by the steepness of the graph line. It can be seen in Figure 6.2(*b*) that in both networks average segment length increases more slowly between third-order and fourth-order streams than between streams of lower orders.

In making a comparison between the detailed stream patterns of different drainage basins it is useful to begin by constructing graphs which relate stream number and stream length to stream order, like those given in Figure 6.2(*b*).

6.2 Transport networks

TOPOLOGY (GRAPH THEORY)

The study of transport networks in geography has relied a good deal on *topology* or *graph theory*, which is a kind of geometry. The networks have been regarded as being made up of *vertices* (points or nodes) and lines (edges, arcs or links). Vertices are situated at the meeting points of two or more edges. Some are settlements but others are just junctions of edges. When several edges occur in succession linking a number of vertices, this is termed a *path*. A *circuit* or *closed path* is a path that begins at any vertex and returns to that vertex without crossing its own route or passing along the same edge twice. A *fundamental circuit* is one that does not enclose other circuits.

Figure 6.3(*a*) is a map of the A class roads and some of the settlements on the Isle of Wight. In Figure 6.3(*b*) the network has been transformed to a topological equivalent. The main change is that the roads have been straightened so as to run directly from settlement to settlement. Each of the settlements has become a vertex and three vertices have

TECHNIQUES USED IN THE ANALYSIS OF NETWORKS 85

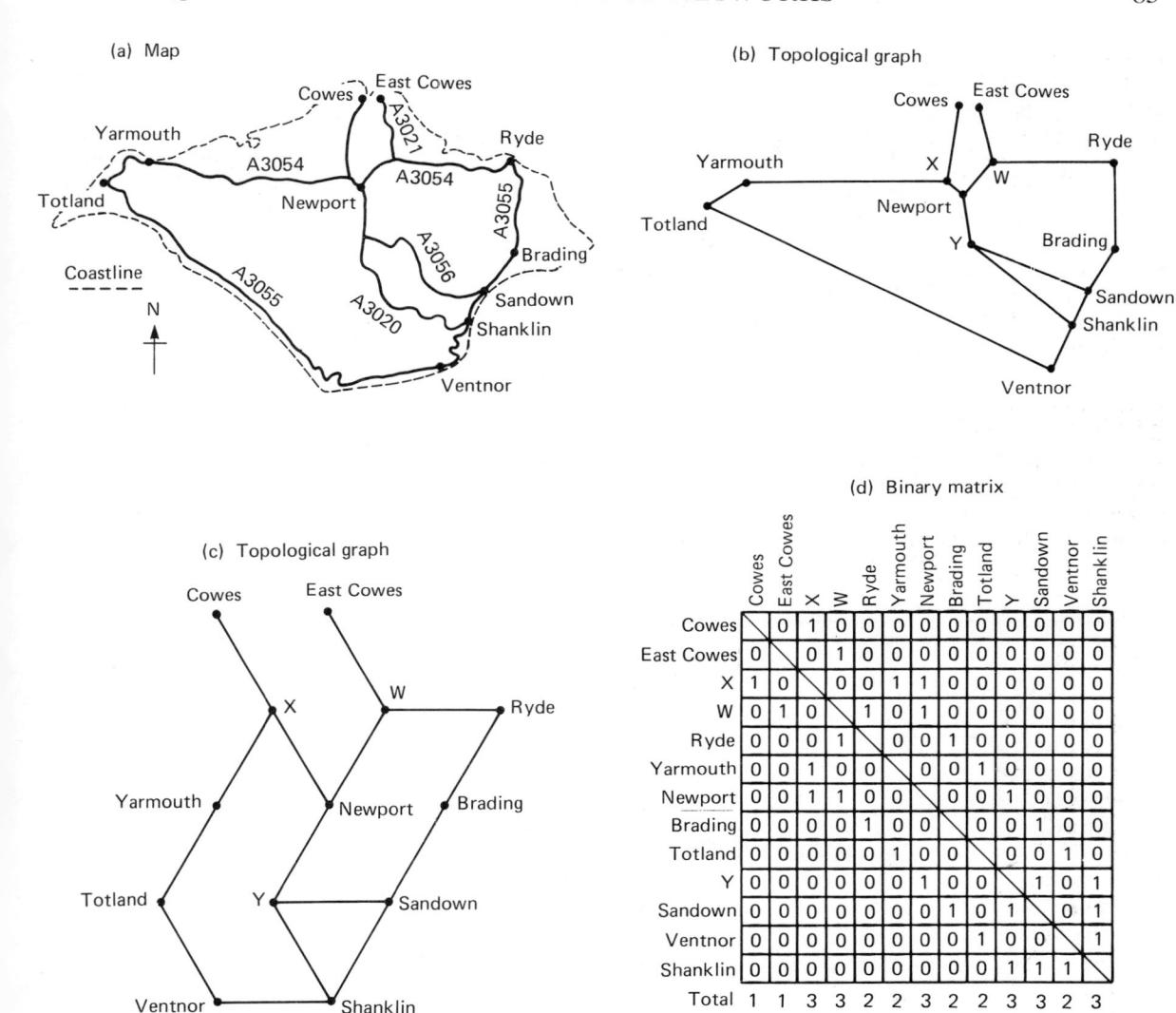

Fig. 6.3 The A class road network of the Isle of Wight

appeared (W, X and Y) at road junctions to the north and south of Newport. On this map the settlements and the road junctions are still in the same places as in Figure 6.3(a). Figure 6.3(c) is also a topological representation of the original map. Here the edges have all been made the same length but the relationships between the vertices and the edges are still topologically similar, and they are very much easier to see and understand.

Figure 6.3(d) is a *binary connectivity matrix* which summarises the information provided in Figure 6.3(a). It shows which pairs of vertices are linked directly with each other by a single edge. It is called 'binary' because it indicates only whether there is a direct link (1) or not (0). From this matrix we can see that Cowes is linked directly to vertex X but not to any other. It also shows, in the column totals, the number of other vertices to which each vertex is directly linked. For example, vertex X is linked directly to three others (Cowes, Yarmouth and Newport). X, W, Newport, Y, Sandown and Shanklin are 3-nodes linked directly by single edges to three other vertices. In Figure 6.3(c) the path from Y to Newport to W to Ryde to Brading to Sandown to Shanklin and back to Y is a circuit. There are three fundamental circuits. These are: Y – Shanklin – Sandown – Y; Sandown – Y – Newport – W – Ryde – Brading – Sandown; and X – Newport – Y – Shanklin – Ventnor – Totland – Yarmouth – X.

CONNECTIVITY

Various indices can be used to describe the *connectivity* of a transport network. The term 'connectivity' means the extent to which the vertices of the network are connected to one another. Explanation here is restricted to *planar graphs* which represent networks that are all in the same plane, so that edges can only cross one another at vertices. Most road networks are of this type but of course air route networks and parts of motorway systems are not confined to a single plane. The minimum number of edges needed to link together all the vertices of a network is $v - 1$, where v is the number of vertices. This minimum number will produce a network in the form of a 'tree'. Neither botanical nor topological trees have circuits. In Figure 6.4(a) there are three edges, one less than the number of vertices (i.e. $v - 1$). If one more edge is inserted along one of the dashed lines, a circuit is created. The maximum number of edges in a planar network is $3(v-2)$. This is shown in Figure 6.4(b). Here there are four vertices so there can be $3(4-2)$ (i.e. 6) edges.

(a) Tree

e = number of edges
v = number of vertices

$e = v - 1$
$= 4 - 1$
$= \underline{3}$

(b) Maximum number of edges

Maximum value of e
$= 3(v - 2)$
$= 3(4 - 2)$
$= 3 \times 2$
$= \underline{6}$

(c) Beta index

Beta index $= \dfrac{e}{v}$
$= \dfrac{3}{4}$

(d) Cyclomatic number

Cyclomatic number $= e - v + 1$
$= 5 - 4 + 1$
$= \underline{2}$

Beta index $= \dfrac{e}{v}$
$= \dfrac{4}{4}$
$= \underline{1}$

(e) Alpha index

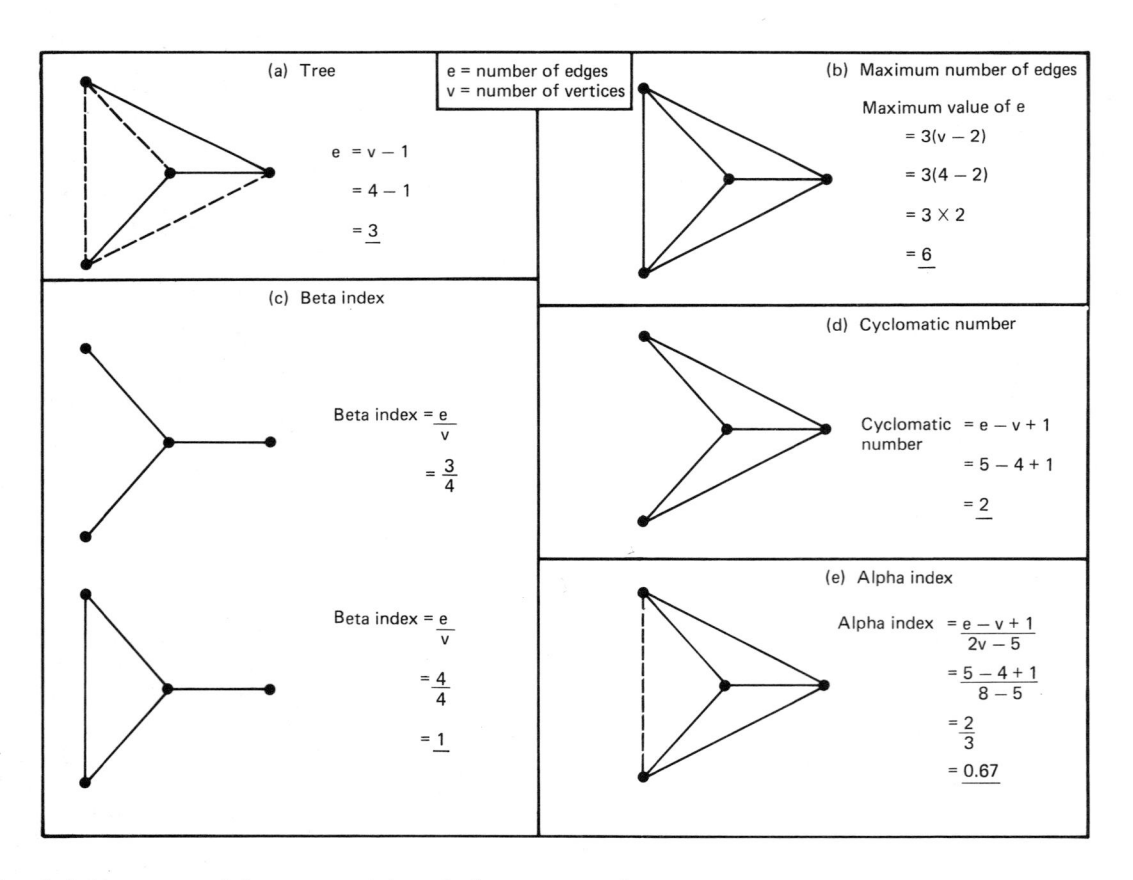

Alpha index $= \dfrac{e - v + 1}{2v - 5}$
$= \dfrac{5 - 4 + 1}{8 - 5}$
$= \dfrac{2}{3}$
$= \underline{0.67}$

Fig. 6.4 Measures of the connectivity of planar networks

The Beta index, the cyclomatic number and the Alpha index

The connectivity of a network is a measure of the completeness of the links between the various vertices. There are three simple measures of the connectivity of a planar network. The simplest of all is the *Beta index* which is calculated by dividing the number of edges (e) by the number of vertices (v) (Fig. 6.4(c)). If the Beta index is 1 there must be a circuit in the network. If the network is in the form of a tree the Beta index is less than 1. The *cyclomatic number* indicates the number of fundamental circuits in the network. It is calculated by using the expression $e - v + 1$ for a simple connected graph. In Figure 6.4(d) there are two fundamental circuits, as predicted by the cyclomatic number. If however the network is divided into two or more separate subgraphs the cyclomatic number is calculated by: $e - v +$ the number of subgraphs, which is of course the general formula applicable also to connected networks. Usually it is easier to find the cyclomatic number by simply counting the number of circuits instead of using the formula. The cyclomatic number is too simple to be a good indicator of connectivity. It is possible for two networks to have the same cyclomatic number and still have very different degrees of connectivity.

Our last measure of connectivity is the *Alpha index*. This is calculated by dividing the cyclomatic number (the number of fundamental circuits) by $2v - 5$ (the maximum possible number of fundamental circuits for the network concerned). A value of 0 for the Alpha index indicates that the network has no fundamental circuit at all. The maximum value of the index is 1 (or 100%). This indicates that the network has the maximum possible number of fundamental circuits and consequently the maximum degree of connectivity. The graph in Figure 6.4(e) has an Alpha index of 0.67. One more edge along the dashed line would increase the connectivity to the maximum possible value.

ACCESSIBILITY

The edges of a network also provide *accessibility* for each vertex and also for the network as a whole. A *shortest-path matrix* shows the topological distance (measured in terms of edges) between each pair of vertices. In the networks of Figure 6.5(a) it is assumed that movement is possible in either direction along any edge. The degree of accessibility provided by a network partly depends on the network's connectivity. If a network has a large number of edges in relation to its vertices it is more likely that there will be fairly direct routes between the vertices and therefore a high level of accessibility. But networks of the same connectivity do not necessarily provide the same degree of accessibility. The three networks in Figure 6.5(a) are all equal in terms of connectivity. All have six edges and six vertices. Hence all have one fundamental circuit ($e-v+1 = 6-6+1 = 1$). In each case however the arrangement of the linkages between the vertices is different and this causes each network to provide to provide a different degree of accessibility. Accompanying each of the networks is its shortest-path matrix showing the number of edges that are traversed in travelling from each vertex to every other vertex. For example, in Network A two edges separate vertex A from vertex B. These are the edges AC and BC. In the last two columns, to the right of the matrix itself are listed the values of the *Shimbel index* and the *Associated number* for each vertex. In Network A, in five successive journeys from vertex A to vertices B, C, D, E and F, altogether ten edges will have been traversed. The Shimbel index is therefore 10. It is the sum of the values in a *row* of the shortest-path matrix. A's row in the shortest-path matrix also shows that its most distant vertex is D, which is three edges away from A. Vertex A's *Associated number* is therefore 3. The Associated number of a vertex is the number of edges between the vertex concerned and the most distant vertex (topologically) in the network. In short it is the greatest value in the row of the shortest-path matrix. The number of edges in the shortest path between the two most distant (topologically) vertices in a network is called the *diameter* of the network. The diameter of Network A is therefore 3 and the diameter extends from vertex A to vertex D.

The Dispersion index, the Mean Shimbel index and the Mean Associated number

Three indices of the level of accessibility provided by the *network as a whole* are given for the networks in Figure 6.5(a). The *Dispersion*

88 PRACTICAL GEOGRAPHY: PRESENTATION AND ANALYSIS

(a)

Network A

Dispersion index = 52
Mean Shimbel index = 8.7
Mean Associated number = 2.5

From \ To	A	B	C	D	E	F	Shimbel index	Assoc. number
A		2	1	3	2	2	10	3
B	2		1	1	1	2	7	2
C	1	1		2	1	1	6	2
D	3	1	2		2	3	11	3
E	2	1	1	2		2	8	2
F	2	2	1	3	2		10	3

Network B

Dispersion index = 58
Mean Shimbel index = 9.7
Mean Associated number = 3.2

From \ To	A	B	C	D	E	F	Shimbel index	Assoc. number
A		4	1	3	2	2	12	4
B	4		3	1	2	3	13	4
C	1	3		2	1	1	8	3
D	3	1	2		1	2	9	3
E	2	2	1	1		1	7	2
F	2	3	1	2	1		9	3

Network C

Dispersion index = 54
Mean Shimbel index = 9.0
Mean Associated number = 3.0

From \ To	A	B	C	D	E	F	Shimbel index	Assoc. number
A		1	1	2	3	2	9	3
B	1		2	1	2	3	9	3
C	1	2		3	2	1	9	3
D	2	1	3		1	2	9	3
E	3	2	2	1		1	9	3
F	2	3	1	2	1		9	3

(b)

Slightly modified Network A

Dispersion index = 58
Mean Shimbel index = 9.7
Mean Associated number = 2.8

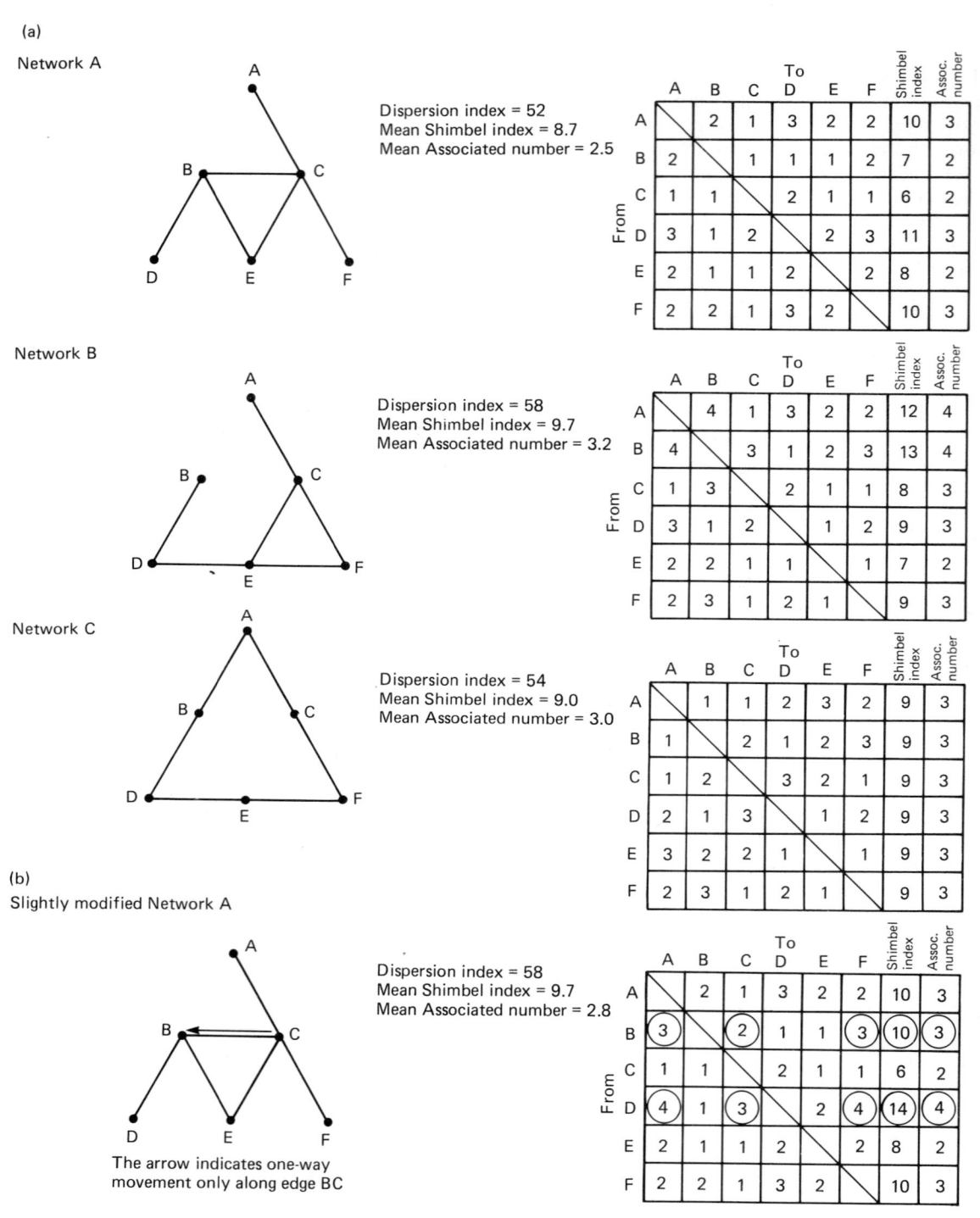

The arrow indicates one-way movement only along edge BC

From \ To	A	B	C	D	E	F	Shimbel index	Assoc. number
A		2	1	3	2	2	10	3
B	(3)		(2)	1	1	(3)	(10)	(3)
C	1	1		2	1	1	6	2
D	(4)	1	(3)		2	(4)	(14)	(4)
E	2	1	1	2		2	8	2
F	2	2	1	3	2		10	3

Fig. 6.5 Measures of accessibility in planar networks

TECHNIQUES USED IN THE ANALYSIS OF NETWORKS

index is the sum of all the Shimbel indices and also the sum of all the inter-vertex distances in the matrix. The *Mean Shimbel index* is the average of all the Shimbel indices. The *Mean Associated number* is the average of all the vertices' Associated numbers.

Of the three networks in Figure 6.5(*a*) it is clear that Network A provides the greatest degree of accessibility since it has the *lowest* values for the Dispersion index, the Mean Shimbel index and the Mean Associated number. Network B provides the lowest level of accessibility mainly because of the high Associated numbers and Shimbel indices of vertices A and B. The circuit is not centrally located within the network and the diameter is

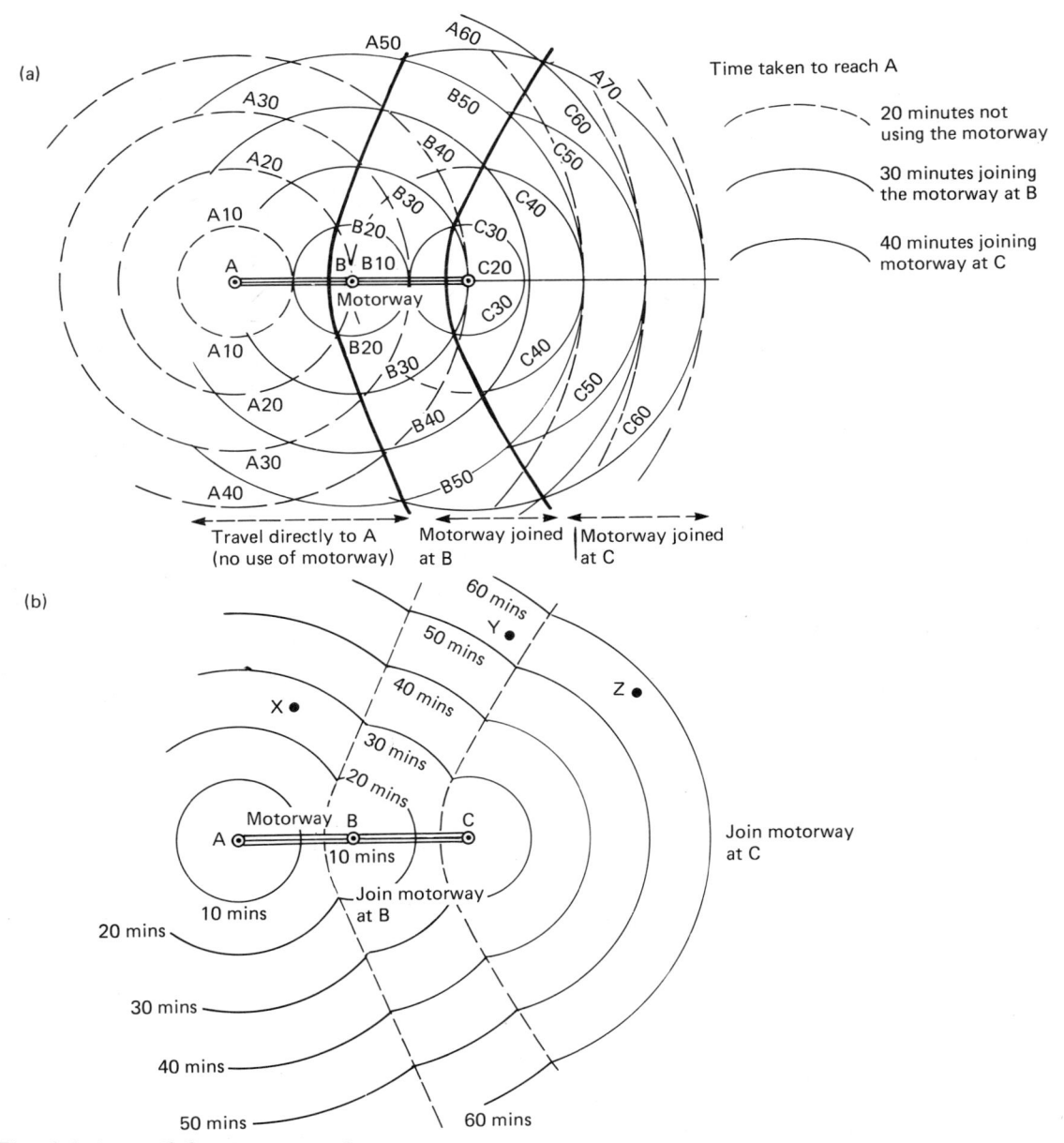

Fig. 6.6 Accessibility in terms of time

4, larger than that of either network A or network C. Figure 6.5(b) shows the effect on Network A (Fig. 6.5(a)) of restricting one edge (BC) to movement in only one direction. This network is now described as a *directed graph*. The general level of accessibility in the network has clearly decreased. The dispersion index has increased from 52 to 58; the mean Shimbel index has increased from 8.7 to 9.7 and the mean Associated number has increased from 2.5 to 2.8. In fact the accessibility of the network has decreased almost to the level of Network B in Figure 6.5(a). Another change that has taken place is that the shortest path matrix is no longer symmetrical in relation to the blank diagonal. For example, in Network A (Fig. 6.5(a)) the topological distance both from A to B and from B to A is two edges. In the directed network however the distance from A to B is two edges but that from B to A is three edges. This is because to travel from B to A it is necessary to travel via E in order to avoid the one-way edge (BC).

Time accessibility

It is also possible to consider accessibility in terms of the *time taken* to travel along the major routes of the network and also across the intervening areas provided that certain simplifying assumptions are adopted. Figure 6.6(a) shows the location of a motorway which has access and exit points at A, B and C. The non-motorway road network is not shown but it is assumed that vehicles are able to travel at such a speed along ordinary roads that it is possible to travel from each of the dashed circles to A in the number of minutes indicated. The other assumption is that it is possible to travel twice as fast along the motorway as across the rest of the area. For example, using the ordinary road network, a vehicle can travel from the dashed circle labelled A10 (Fig.6.6(a)) to A in 10 minutes, but along the motorway it is possible to travel to A from B, which is twice as far away, in 10 minutes. A journey along the motorway from C to A takes 20 minutes (C20). By the ordinary road system it takes 40 minutes (dashed line A40). Figure 6.6(b) shows the pattern of time-accessibility that results from Figure 6.6(a). To reach A in the minimum possible time a vehicle starting at X should not use the motorway. One starting at Y should join the motorway at B and the journey time would be 45 minutes compared with nearly 50 minutes using the non-motorway road network. A vehicle starting at Z can reach A in about 56 minutes compared with over 70 minutes if the motorway were not used.

Exercises

1. (a) Use the sketch map below of a drainage basin to calculate the following descriptive indices:
   - (i) Its form factor;
   - (ii) Its shape index;
   - (iii) Its basin relief;
   - (iv) Its drainage density.
   (b) Study the pattern of drainage channels in the basin, then compile tables showing:
   - (i) The number of stream segments in each stream order;
   - (ii) The bifurcation ratio between successive stream orders;
   - (iii) The length ratio between successive stream orders.
   (c) On semi-logarithmic graph paper plot:
   - (i) The number of stream segments against stream order;
   - (ii) The mean length of stream segments against stream order.
   (d) Compare the stream network shown below with that shown in Figure 6.2(a) Network B.

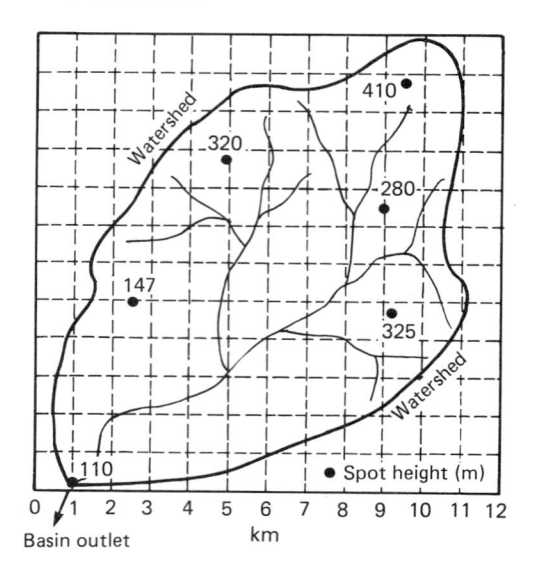

TECHNIQUES USED IN THE ANALYSIS OF NETWORKS

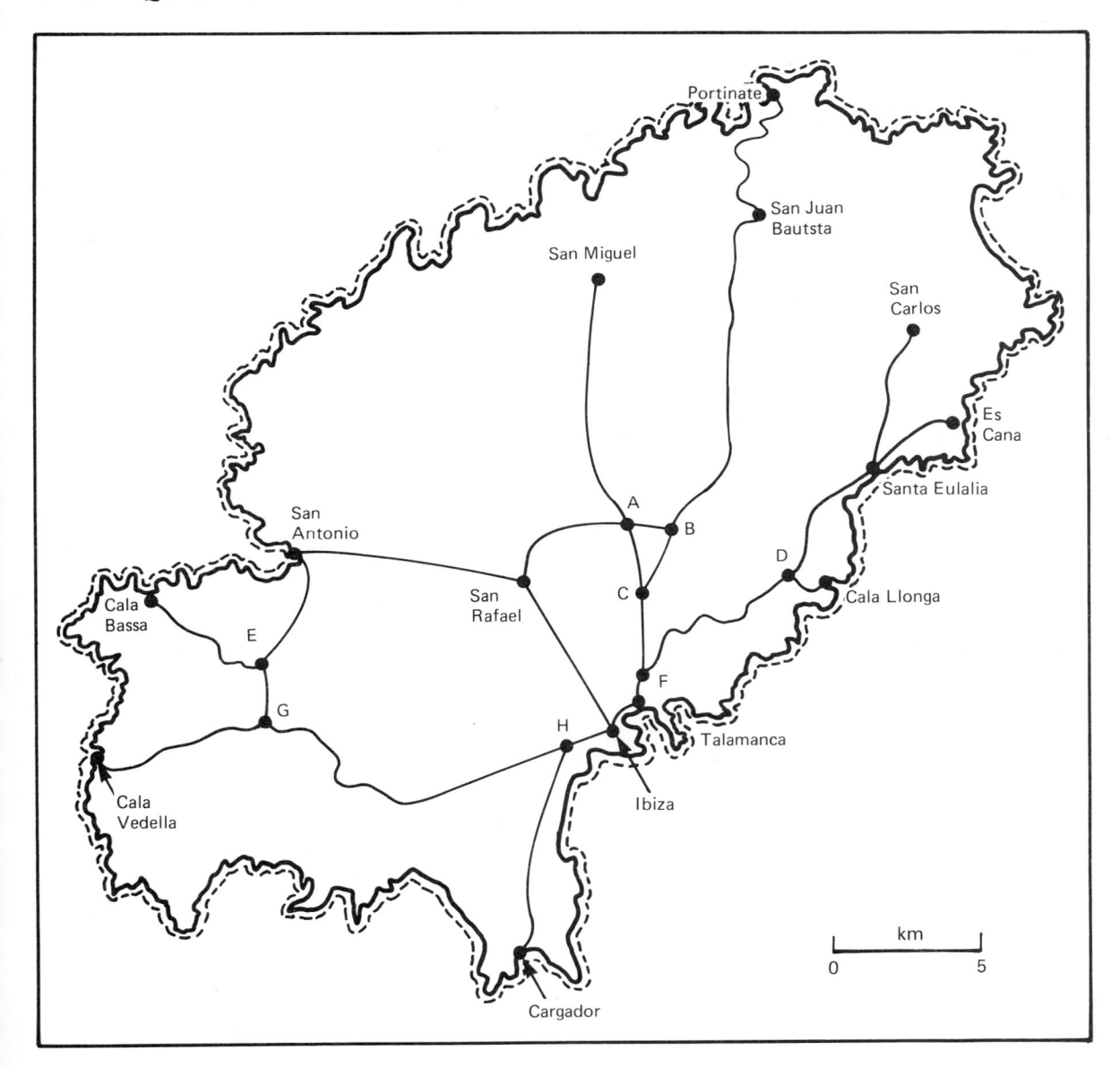

2. The map shows the main features of the road network of the island of Ibiza in the Mediterranean Sea.
   (a) Transform the main features of the road network of Ibiza into its topological equivalent (as shown in Fig. 6.3(c) for the Isle of Wight).
   (b) Construct a binary connectivity matrix, as in Figure 6.3(d), for the Ibiza road network.
   (c) For the Ibiza road network calculate:
   (i) The Beta index; (ii) The cyclomatic number; (iii) The Alpha index.
   (d) Construct a shortest-path matrix for the Ibiza road network and calculate its Dispersion index, its mean Shimbel index and its Associated number (page 00).
   (e) Write a comparison between the road networks of Ibiza and the Isle of Wight (Fig. 6.3).

7 Specialisation—the Location Quotient

The location quotient is an index which can be used to assess the share of several smaller regions in a spatial distribution which extends over a much larger region. For example, it can show the shares of separate counties in a nationwide distribution or the share of separate countries in a distribution extending across a continent. It can indicate the extent to which specialisation occurs in any of these smaller regions. Figure 7.1(a) shows a simple specimen

(a)

Each dot represents a population of 2000

Total area = 50 sq km

Region	Area (sq km)	Percentage of total area
A	10	20
B	5	10
C	15	30
D	15	30
E	5	10
	50	100

Total population = 100 000 (50 dots)

Region	Population (thousands)	Percentage of total population
A	40	40
B	10	10
C	20	20
D	10	10
E	20	20
	100	100

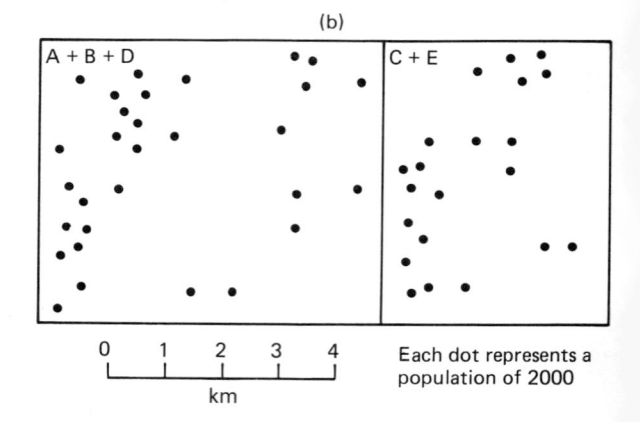

Region	Percentage of total area	Percentage of total population
A	20	40
B	10	10
C	30	20
D	30	10
E	10	20
	100	100

Location quotient
$$\left(\frac{\%\ of\ total\ population}{\%\ of\ total\ area}\right)$$

A 2.0
B 1.0
C 0.67
D 0.33
E 2.0

(b)

Each dot represents a population of 2000

92

SPECIALISATION—THE LOCATION QUOTIENT

Region	Total area = 50 sq km Area (sq km)	Percentage of total area
A+B+D	30	60
C+E	20	40
	50	100

Region	Total population = 100 000 (50 dots) Population (thousands)	Percentage of total population
A+B+D	60	60
C+E	40	40
	100	100

Region	Population (thousands)	Percentage of total population
A+B+D	60	60
C+E	40	40
	100	100

Location quotient
$$\left(\frac{\% \text{ of total population}}{\% \text{ of total area}}\right)$$

A+B+D 1.0
C+E   1.0

Fig. 7.1 The calculation of the location quotient

distribution of population. The diagram here represents an area divided into five separate regions (A to E). These regions may be counties within a country or areas within a city or conurbation. The aim is to discover whether the population totals of the various regions correspond with their sizes. The map in Figure 7.1(*a*) suggests that regions A and E are the most densely populated and that region D is the most sparsely populated.

Calculating the location quotient
To calculate the location quotient it is necessary first of all to calculate the percentage of the total area that is occupied by each region. Regions C and D are the largest, each occupying 30% of the total area. In contrast regions B and E are the smallest, each occupying only 10% of the total area. Area A occupies 20%. Next we have to calculate for each region its percentage share of the total population of the area. It can be seen that region A has 40% of the total population, C and E each have 20% and B and D have only 10% each. These area and population values are tabulated in Figure 7.1(*a*).

The location quotient for each region can now be calculated simply by dividing the percentage of its total population by the percentage of its total area, as shown in Figure 7.1(*a*).

Regions A and E have a location quotient of 2.0. This means that in these regions population is twice as densely concentrated as in the area as a whole. In A for example 40% of the total population is concentrated in 20% of the total area, and in E 20% of the total population is concentrated in 10% of the total area. In region B the location quotient is 1.0. Here 10% of the total population is found in 10% of the total area. This gives a population concentration equal to that of the area as a whole. In C and D population is more thinly spread than in the area as a whole, particularly in D which has only one-third of the density of the whole area.

It is clear therefore that a high value of the location quotient is associated with a high concentration of population. In this example there appeared a wide range of location quotients. If, however, the area is subdivided into fewer, but larger regions, this variation tends to be much reduced. In Figure 7.1(*b*) the overall population distribution is the same as in Figure 7.1(*a*) but there are only two regions instead of five. Regions A, B and D have been combined into a single region, and so have regions C and E. The result is that 'specialisation' has declined greatly. Both of these regions have a location quotient of 1.0 which means that the population concentration is equal to that of the area as a whole.

Figure 7.1 is only one of many cases in which the location quotient is relevant. Instead of considering population distribution over an area, as in Figure 7.1, it is possible to use a number of other spatial variables. These include employment in various industries and production of various crops. These need not always be related to area. They can also be related, for example, to population numbers.

Suppose, for example, a certain area of England produced 10% of the country's wheat production and this area occupied only 5% of the area of England, then its location quotient would be:

$$LQ = \frac{\text{Percentage of wheat production}}{\text{Percentage of area}} = \frac{10}{5} = 2.0$$

In other words the area produced twice as much wheat as would be the case if wheat production was uniformly distributed over England. Another area with only 10% of the country's industrial workers might produce 30 per cent of the county's motor cars. In this case the location quotient would be:

$$LQ = \frac{\text{Percentage of motor car production}}{\text{Percentage of industrial workers}} = \frac{30}{10} = 3.0$$

In other words the area is producing three times the number of motor cars as it would be if motor car production were evenly distributed in relation to industrial workers.

(a) Calculate a location quotient relating urban population to total population for each of the areas listed, in 1959 and 1979.
(b) Referring to the location quotients you have calculated, write a description of the changes that took place in the distribution of urban population in the USSR between 1959 and 1979.

Exercise

1. The table below shows the changes that took place in the total population and the urban population of the various republics and economic regions of the USSR between 1959 and 1979.

	Population (millions)			
	Total population		Urban population	
	1959	1979	1959	1979
Russian Republic	117.5	137.6	61.6	95.4
Ukrainian Republic	41.9	49.8	19.1	30.5
Belorussian Republic	8.1	9.6	2.5	5.3
Baltic Republics	6.0	7.4	2.9	4.8
Moldavian Republic	2.9	3.9	0.6	1.6
Transcaucasus	9.5	14.1	4.4	7.8
Kazakh Republic	9.3	14.7	4.1	7.9
Central Asia	13.7	25.5	4.8	10.4

8 The Spearman Rank Correlation Coefficient

CORRELATION

The relationships between two geographical variables can be studied by using the technique of correlation. This involves the comparison of one set of data with the other set and the calculation of a *correlation coefficient* that summarises the relationship between the two variables. Spearman's technique of *rank correlation* is one of the most useful of these techniques and it is also quite easy to use. It is based upon the *ranks* of the individual values of the two variables rather than the values themselves. It can even be used in cases in which the precise values are not known provided that the rank order is known.

Calculating a rank correlation coefficient

In calculating a rank correlation coefficient it is first of all necessary to arrange the individual values of each variable in rank order. The highest value of each variable is given the rank of 1 and successive lower values are given the ranks of 2, 3, 4, 5 etc. An example is given below.

Variable X
Actual values: 12 10 9 7 6 4 3 1
Ranks: 1 2 3 4 5 6 7 8

Sometimes two or more of the values are equal (i.e. tied). In this case each of the values is given the average of the ranks which would otherwise have been allocated. This is illustrated in the case of variable Y below.

Variable Y
Actual values: 9 8   8   7 4 3   3   1
Ranks: 1 2.5 2.5 4 5 6.5 6.5 8

For example, variable Y has tied values of 8 occupying the 2nd and 3rd ranks. Each of these

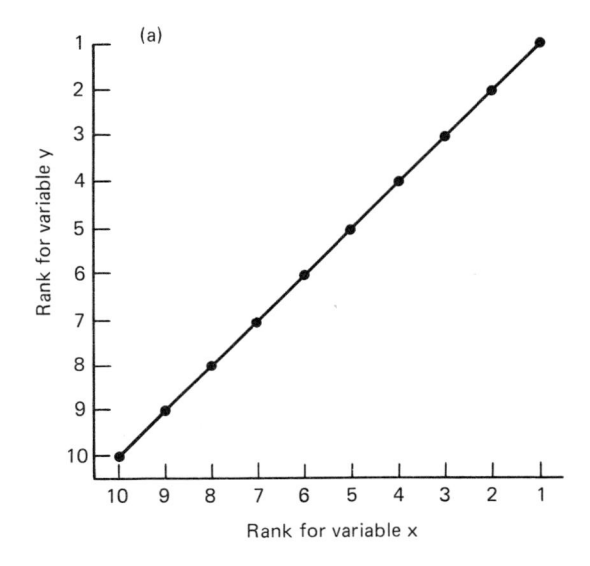

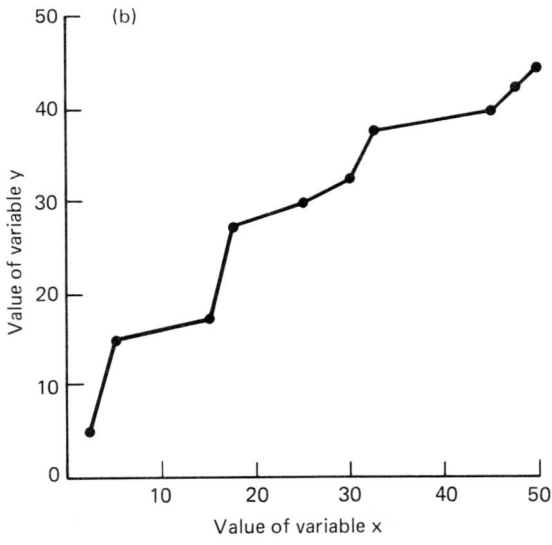

Fig. 8.1 The effect of ranking on the relationship between two variables

therefore is given the rank of 2.5 i.e. $\frac{2+3}{2}$. Similarly tied values of 3 occupy the 6th and 7th ranks, so each of these is given the rank of 6.5 i.e. $\frac{6+7}{2}$. This rule still applies if there are three or more tied values.

This ranking process means that a certain amount of information which was present in the actual values is lost in the process of ranking. For example, in variable Y, the actual values of 7 and 4 differ by three units, but their difference in rank is only 1.0 (5−4). On the other hand, the actual values of 9 and 8 differ by only one unit whereas their rank difference is 1.5.

Figure 8.1(a) shows an example of a perfect rank correlation between variables X and Y. Ranks are identical for both of variables X and Y. This however does not mean that the actual values for variable X are exactly equal to the corresponding values for variable Y. The actual values for each variable could well be as shown in Figure 8.1(b). Inevitably therefore the Spearman Correlation Coefficient involves a certain degree of inaccuracy through the use of ranks instead of actual values.

In using rank correlation for the purpose of an actual investigation of the relationship between two sets of variables it is usual to follow the procedure described below.

The first step is to formulate the *null hypothesis* (often abbreviated to H_o). This is simply a statement that the two sets of ranks are quite independent of one another and there is no systematic relationship between them. Unless we can prove otherwise by the level of significance of the correlation coefficient, this null hypothesis holds good. Next the *alternative hypothesis* H_1 is formulated. This is a statement of the way in which we *think* that the two sets of data are related to each other. The alternative hypothesis can be a statement that the two sets of ranks are positively correlated or that they are negatively correlated. Such an arrangement is termed a one-tailed test. The rank correlation coefficient is then calculated. If this proves to have a very low value we have no alternative but to accept the null hypothesis that the two sets of ranks are not systematically related to each other. On the other hand, if the coefficient has a high value we may be justified in rejecting the null hypothesis and accepting the alternative hypothesis. Whether or not this occurs depends upon the level of *significance* of the correlation coefficient.

The simple logic of this somewhat laborious process is that we are not entitled to assume that there is a certain relationship between the two sets of ranks until we can actually prove that one exists. The *significance level* of a correlation coefficient is the probability that the correlation could have occurred entirely by chance. If the correlation had in fact occurred by chance we should not be justified in rejecting the null hypothesis and accepting the alternative hypothesis. Figure 8.2 shows the relationships between the rank correlation coefficient, the number of pairs in the data and two levels of significance (0.01 and 0.05). A significance level of 0.01 means that there is 1 chance in 100 (1%) of our having rejected the null hypothesis wrongly (i.e. of the correlation having occurred by chance). Such a low value of significance is clearly associated in Figure 8.2 with a large number of pairs in the data and a high value of the correlation coefficient. A significance level of 0.05 (a 5% or 1 in 20 chance of the correlation having occurred by chance) is attainable with fewer pairs in the data or with a lower value of the correlation coefficient. The significance level of 0.05 is generally regarded as the greatest risk that should be taken of the correlation having occurred by chance.

Calculating Spearman's Rank Correlation Coefficient

Spearman's Rank Correlation Coefficient is easy to calculate. A specimen calculation is given below.

Ranks of variable X	1	2	3	4	5	6	7	8
Ranks of variable Y	2	4	1	5	8	3	6	7
d (difference between each pair of ranks)	1	2	2	1	3	3	1	1
d^2	1	4	4	1	9	9	1	1	$\Sigma d^2 = 30$

$$\text{Rank Correlation Coefficient} = 1 - \frac{6\Sigma d^2}{n^3 - n}$$

$$= 1 - \frac{180}{504}$$

$$= 1 - 0.357$$

$$= \underline{0.643}$$

THE SPEARMAN RANK CORRELATION COEFFICIENT

First of all variable X is ranked beginning at its largest value. Next the corresponding ranks of variable Y are listed. Then d (the rank differences beween those of variable X and the correponding values of variable Y) is calculated for each pair of ranks. This d value is then squared for each pair of ranks. Next $n^3 - n$ is calculated (n being the number of pairs of ranks, in this case 8). These calculated values are then inserted in the general formula for the calculation of the rank correlation coefficient, as shown above.

To test the null hypothesis that the ranks of variables X and Y are independent we turn to Figure 8.2. The correlation coefficient is 0.643 and the number of pairs is 8. We mark this location on the graph and we find that it falls on the curve showing a significance level of 0.05. This value just allows us to reject the null hypothesis that variables X and Y are independent and allows us to accept the alternative hypothesis that the ranks of the two variables are positively correlated. It is clear however that the correlation is not particular strong. Even a slight reduction in the correlation coefficient would have compelled us to accept the null hypothesis. Figure 8.2 also shows that if a correlation coefficient of 0.643 had been obtained for a pair of variables with 16 pairs of ranks the correlation would have been much stronger, with a significance level less than 0.01 (Fig. 8.2).

It is important to remember that a very high correlation coefficient between two variables, X and Y, does not necessarily mean that the variations in variable X have *caused* the variations in variable Y, or that the variations in variable Y have *caused* the variations in variable X, though it is *possible* that either of these has occurred. It is also possible however that

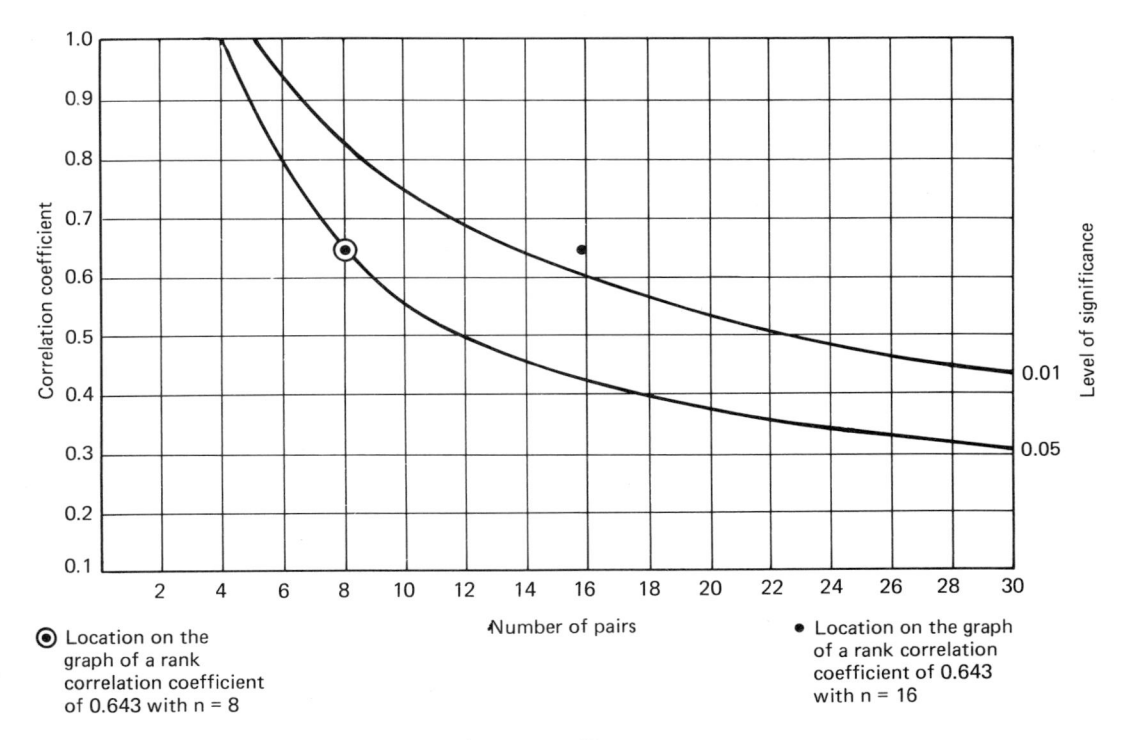

Fig. 8.2 Critical values of the rank correlation coefficient

◉ Location on the graph of a rank correlation coefficient of 0.643 with n = 8

● Location on the graph of a rank correlation coefficient of 0.643 with n = 16 variations in variable Z, which has not been involved in the correlation process, have influenced both of variables X and Y. For example, variables X and Y could be mean January and mean July temperatures, and variable Z might be latitude.

A few simple relationships relevant to the Spearman Rank Correlation Coefficient are easy to understand. If the two variables have exactly the same rankings the correlation coefficient is always 1.0, a perfectly positive correlation. This is because the value of d in the formula would then be zero, so 1 minus zero equals 1.0. If one of the variables has rankings that are in exactly the reverse order from the other then the correlation coefficient has a value of -1.0, representing a perfect negative correlation. If the ranks of one of the variables are all equal then the correlation coefficient is $+0.5$. Hence it is undesirable to have too many tied ranks in the data used for correlation.

(a) Give a concise description of the distribution of malaria in Bangladesh in each of the two time periods.
(b) Use Spearman's Rank Correlation technique to examine the validity of the null hypothesis that there is no systematic relationship in the incidence of malaria between the two time periods. The alternative hypothesis is that there is a positive relationship.

Exercise

The table below shows the incidence of malaria (per 100 000 persons) in the various districts of Bangladesh in 1968 – 71 and 1972 – 77.

District	1968 – 71	1972 – 77
Dhaka	1.8	12.5
Tangail	1.0	4.0
Mymensingh	10.0	20.3
Faridpur	4.5	30.3
Chittagong	43.5	176.5
Chittagong Hill Tracts	47.3	1171.2
Noakhali	2.3	18.7
Comilla	7.8	24.6
Sylhet	27.2	77.8
Rajshahi	0.5	1.5
Pabna	0.8	2.0
Bogra	1.0	1.8
Rangpur	10.0	8.2
Dinajpur	0.8	1.8
Jessore	1.8	1.5
Kushtia	0.8	1.7
Barisal	6.8	10.7
Patuakhali	27.3	6.3
Khulna	4.5	5.3

9 Weather Maps

To understand the weather that is depicted on a weather map it is necessary to understand the various processes that have created it. Much of the weather of Western Europe is the result of interaction between *air masses*.

AIR MASSES

An air mass is a large volume of air whose temperature and humidity have become almost uniform through remaining in a 'source region' for a considerable time.

Figure 9.1 shows some of the source regions of some of the air masses which affect the British Isles. These source regions are classified first into 'continental' (c) and 'maritime' (m) and then into 'tropical' (T), 'polar' (P) and 'arctic' (A).

Through the year one or other of these air masses advances towards the British Isles and

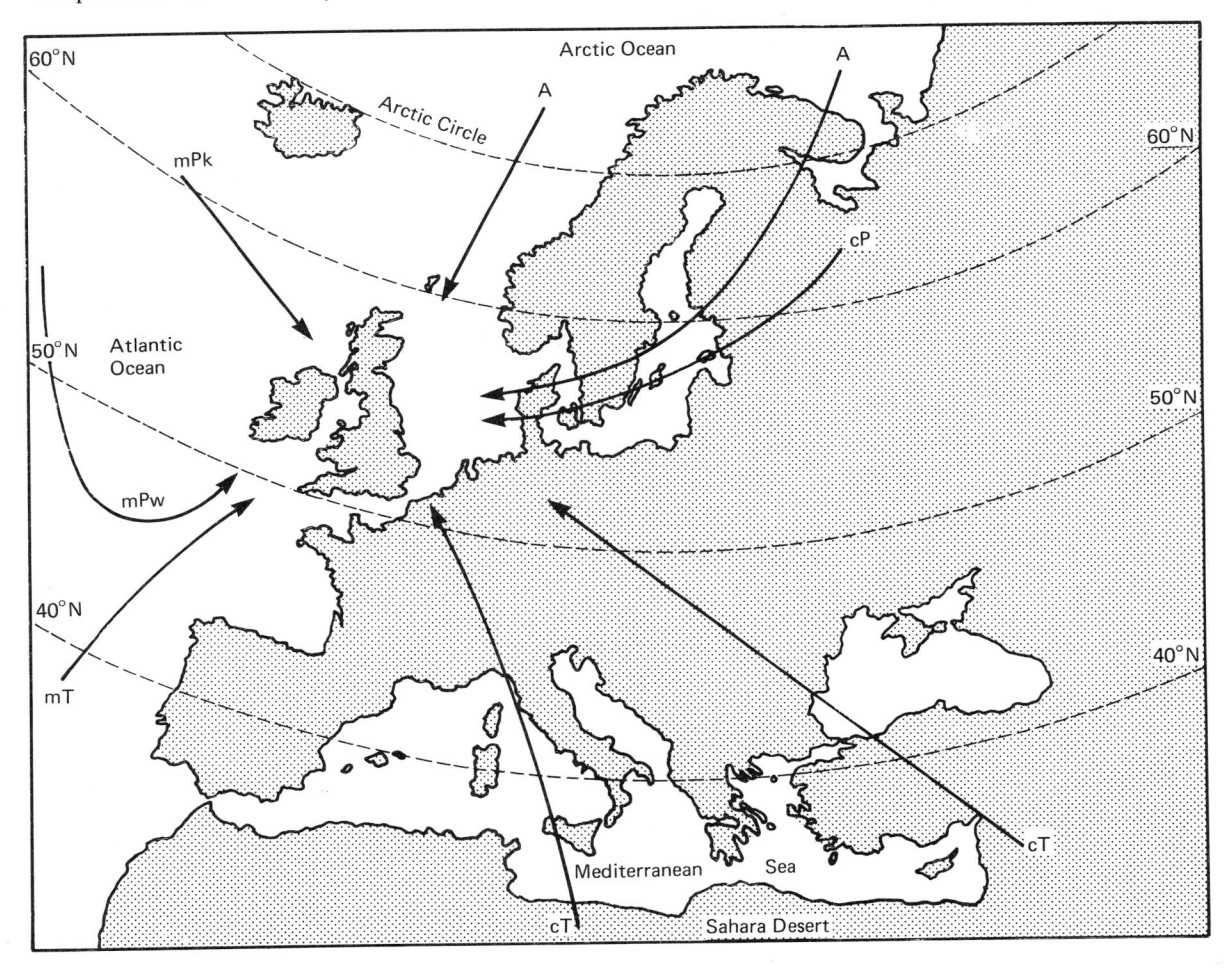

Fig. 9.1 Air masses effecting the British Isles

establishes a particular type of weather that is related to the meteorological characteristics of its source area. Our weather is so variable because it is rare for any single air mass to dominate for any considerable length of time, and also because the air masses that influence British weather are of both polar/arctic and tropical origin. Also, sometimes two contrasting air masses can affect Britain at the same time. In this case a *front* may be created between the two air masses where conditions of temperature and humidity change very rapidly.

Arctic and polar air masses

The Arctic air mass (A) (Fig. 9.1) is very cold indeed in winter, particularly if it travels across the cold land mass of Russia and Scandinavia. If, however, it reaches Britain by the oceanic route its lower layers will be warmed to some extent by the sea. Hence it may become unstable and bring snow to Britain. In winter too continental polar (cP) air may bring cold weather to Britain. Snow is likely to fall along the east coast but the west coast is likely to have cold, dry sunny weather. In summer Scandinavia is quite warm so the cP air moving towards Britain is also warm. However its lower layers may be cooled in early summer as it crosses the North Sea. This can result in condensation and the development of low stratus clouds and perhaps fog on the east coast.

The maritime polar air mass has a low temperature in its source region (Fig. 9.1) but, on its way to Britain its lower layers tend to be warmed in winter by the North Atlantic Drift (mPk). This may cause some instability which can result in the formation of cumulus clouds and the occurrence of showers, sometimes of snow, in winter. These snow showers tend to occur mainly in western districts. Few extend very far inland because the cold land surface tends to cool the lower layers of the air mass and hence to damp down convection. In summer however showery weather can spread inland. The air mass tends to become unstable over the land because the relatively warm land surface in summer tends to warm its lower layers and produce a steeper lapse rate of temperature. Sometimes maritime polar air streams follow a more southerly course across the Atlantic Ocean and then swing north-eastwards towards Britain (Fig. 9.1, mPw air). As this air stream moves north-eastwards its lower layers tend to be cooled from below, so it becomes more stable. A shallow inversion can be formed, under which the weather is rather mild and cloudy (low cloud), with occasional drizzle.

Tropical air masses

The 'tropical' air masses (mT and cT) originate in the high pressure systems that tend to occur over the Azores, the Sahara Desert and south-west Asia (Fig. 9.1). The maritime tropical air mass is usually located near the middle of the North Atlantic between Portugal and the United States' coast, at about latitude 30° – 40° north. This air mass is very humid in its lower layers through evaporation of water from the sea. As it moves northwards towards Britain its lower layers come into contact with a progressively cooler sea surface. This causes condensation which can result in the formation of low stratus clouds or even mist and fog at sea level. It can bring considerable cloud and mist to western Britain and in winter this can drift eastwards across Britain. Temperatures in winter will be quite mild. The mT air mass can also give mist and low cloud to the British west coast in summer but, further inland, these tend to break up over the warmer land because an increase in the environmental lapse rate tends to encourage convection. Occasionally this can result in the occurrence of thunderstorms inland. Continental tropical air from the Sahara and south-west Asia can bring very hot weather to Britain in summer and pleasant, mild, dry weather in winter. Sometimes however this very warm, dry air can absorb moisture over the Mediterranean or the Bay of Biscay on its way north and thunderstorms can occur in southern England.

The weather of north-west Europe is also greatly influenced by the boundary zones between contrasting air masses. Air masses with differing temperatures do not readily mix. Instead a fairly clear-cut boundary zone tends to persist between them. This boundary is referred to as a *front*. Fronts and the *depressions* that are associated with them are to a great extent responsible for the changeability of Britain's weather.

WEATHER MAPS

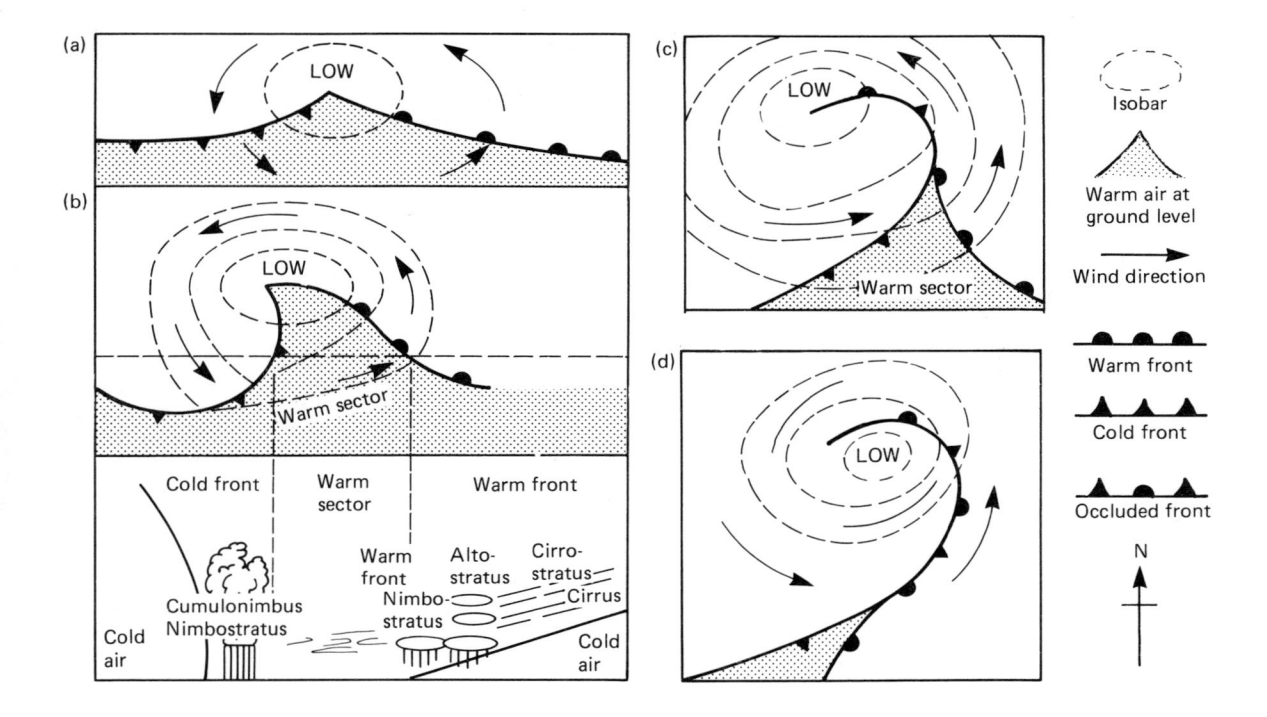

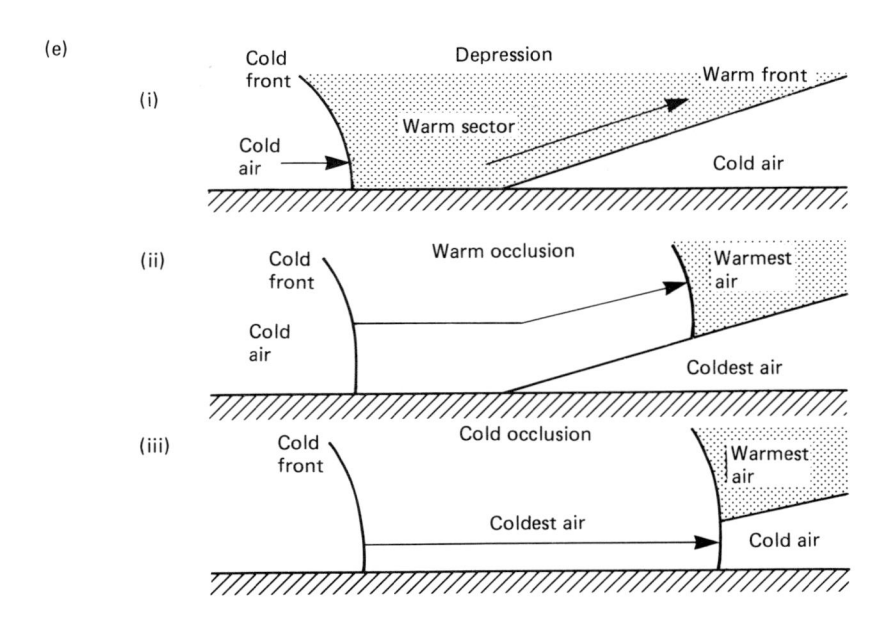

Fig. 9.2 The development of a depression

Depressions

A *depression* is an area of relatively low atmospheric pressure with a generally circular pattern of isobars. It usually occurs at the junction between contrasting air masses, the colder air mass usually lying to the north. A single depression can have a diameter of up to about 2000 km and it usually remains active for about a week. It can originate as a kind of 'wave' that forms on a front separating a mass of relatively warm air from a mass of colder air. Here atmospheric pressure begins to fall (Fig. 9.2(*a*)). The wave then becomes accentuated, atmospheric pressure falls further (the depression deepens), and a distinct warm front and cold front develop (Fig. 9.2(*b*)).

The depression illustrated in Figure 9.2(*b*) as a whole is moving towards the east. The warm front and cold front symbols are placed on the side of the line towards which the front is moving. At the warm front air from the warm sector is rising over the colder air in front of it. At the cold front the colder air to the west is undercutting the warmer air of the warm sector.

Some of the weather conditions typically associated with a depression are illustrated in Figure 9.2(*b*). The first sign of the approach of the depression is the occurrence of streaks of very high cirrus clouds. Next the cloud cover becomes progressively more complete and lower in altitude. The sequence usually is: cirrus, cirrostratus, altostratus and finally, at the lowest level, nimbostratus, from which rain falls. Throughout this cloud sequence barometric pressure has steadily fallen. In the warm sector barometric pressure becomes more constant and the rain often ceases. Soon however the cold front arrives. This is much steeper than the warm front and it can bring very heavy rain for a comparatively short period from cumulonimbus clouds. The rain belt at the cold front of a depression is usually narrower but more vigorous than that of the warm front. Barometric pressure tends to rise after the passage of the cold front.

Figure 9.2(*c*) and (*d*) shows the beginning of the decay of the depression, a process known as *occlusion*. The cold front of the depression overtakes the warm front and the air of the warm sector is raised to a higher level. In a warm occlusion (Fig. 9.2((*e*)(*ii*)) the cold front overtakes the warm front, but the air behind the cold front is warmer than that in advance of the warm front. Hence a *warm occlusion* is created in which the air from behind the cold front rises over the air in front of the warm front. In a *cold occlusion* however the air from behind the cold front is colder than that in advance of the warm front, so the cold front undercuts both the warm sector air and the air in advance of the warm front (Fig. 9.2(*e*)(*iii*)).

In general, a warm occlusion gives weather similar to that of a warm front, with generally stratiform clouds, and a cold occlusion gives weather similar to that of a cold front with cumulus-type clouds. As they move towards Britain from the Atlantic Ocean depressions often occur in 'families'. New depressions come into being on the trailing cold front of the preceding depression. Figure 9.2(*c*) and (*d*) show how a weather map represents the process of occlusion with the cold front 'pinching out' the warm sector. In a procession of Atlantic depressions reaching Britain therefore, those furthest west have probably developed more recently (Fig. 9.2(*a*) and (*b*)) than those nearer to Britain which have tended to become occluded.

Anticyclones

The *anticyclone* is another important feature of Britain's weather. It is an area of high atmospheric pressure that is often even larger than a depression. In an anticyclone pressure increases towards its centre. As in the case of a depression, an anticyclone's isobars are often almost circular. Sometimes however they are elongated into a wedge shape and such a formation is described as a 'ridge' of high pressure. It can occur between pairs of depressions and moves with the depressions, giving alternate spells of rainy and dry weather.

Unlike a depression, the air in an anticyclone is descending and is therefore warmed by compression. Near the ground however there is often a shallow layer of cool air. Hence there can be a marked inversion at a low level. Above the inversion there is usually very little cloud but, beneath the inversion there can be a layer of stratus or stratocumulus, or even fog at ground level. Anticyclones tend to have very light winds. In summer however the sun is more effective in dissipating low cloud and fog, at least during the daytime. Generally anticyclones are associated with dry settled weather. A 'blocking' anticyclone is one that remains stationary for a period of time and impedes the movement of Atlantic depressions towards Western Europe, thus giving an unusually long spell of cold dry weather in winter.

Anticyclones can also influence the direction of movement of air masses. Figure 9.3(a) represents an anticyclone centred over Scandinavia in winter. The clockwise movement of air around this anticyclone brings a stream of very cold continental polar air towards Britain. This air tends to become unstable as it passes over the relatively warm sea and it can bring snow showers to Britain. A situation similar to this occurred in early February 1987, when heavy snow fell in Kent and snow showers extended around the coast from the Wash to Dorset. Maximum daily temperatures were only about 0°C. Further west and north-west however there was dry sunny weather with maximum temperatures rising to 4°C to 6°C.

Figure 9.3(b) shows a contrasting example of the effect of an anticyclone. In this case the area of high pressure is located over France and northern Spain and the Bay of Biscay. The clockwise air circulation draws warm air from the Mediterranean area and also continental tropical air from North Africa. Such a pressure distribution can result in warm, dry weather in Britain.

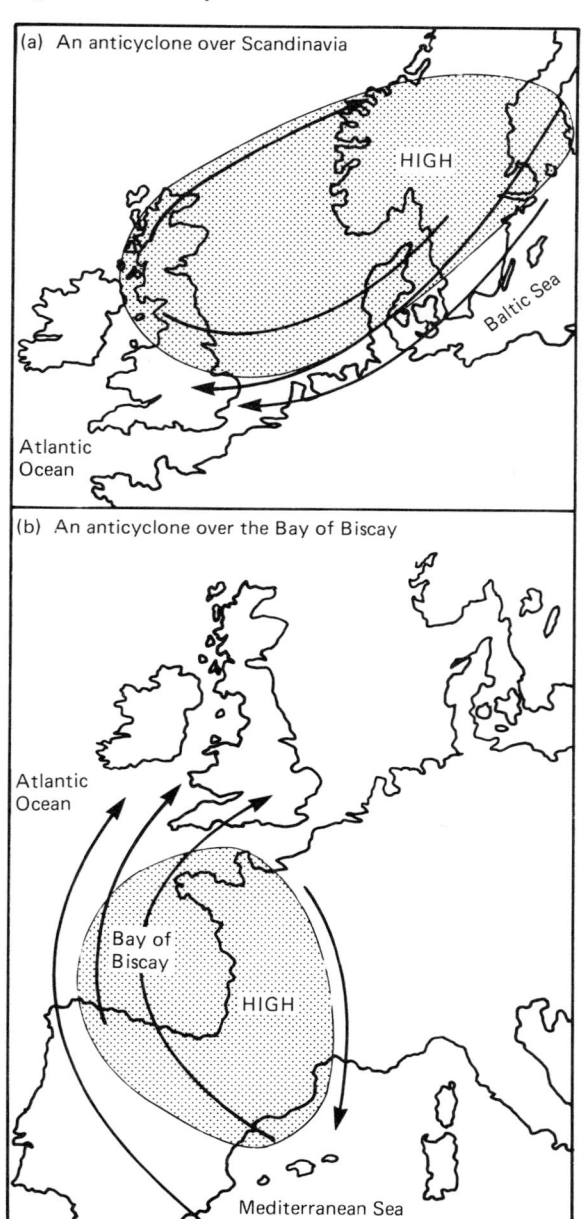

Fig. 9.3 The influence of anticyclones on the movement of air masses

Exercise

Refer to the weather map below.
(a) Give a description and explanation of the variations of weather you would expect to occur over the area of the map.
(b) What changes in the weather would you expect to occur in the next few days?

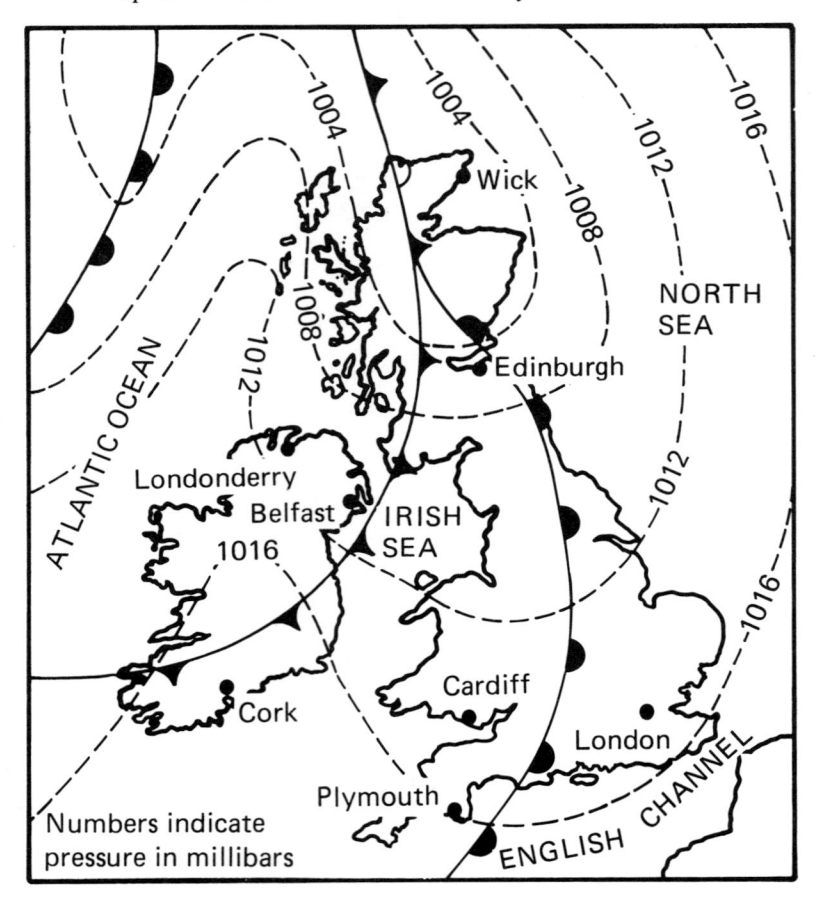

10 Topographical Maps

10.1 Aspects of physical geography

RELIEF

Variations in relief of the land surface are usually depicted on maps by the use of contour lines. Each contour line runs across the map and indicates a constant height of the land surface. Since the land surface varies in height it is rare for these contour lines to be straight lines. Instead they tend to form V-shapes in valleys and on spurs and closed circles indicate isolated hills or depressions (Fig. 10.1(a)). Information about the height of the land surface is also provided by isolated *spot heights* showing the heights at particular points.

A map can give an excellent representation of the relief of an area by means of contour lines each of which represents a different height above sea level. For example, a map may have contour lines for 10m, 20m, 30m and other heights at intervals of 10 m. These contours divide up the land surface into horizontal layers measuring 10 m thick from top to bottom. This means of course that the contours give no information about the detailed shape of the land surface *within* each 10 m zone. Hence, a gently undulating plateau or coastal plain may well have no contours at all over a considerable area yet there can be relief variations of up to 10 m on its surface. Spot heights, however, may give some information.

However, much information about the relief of the land surface can be deduced by studying the *horizontal* distances between the contour lines (horizontal equivalents). For example, a steep slope is indicated by the close packing together of successive contour lines. This is

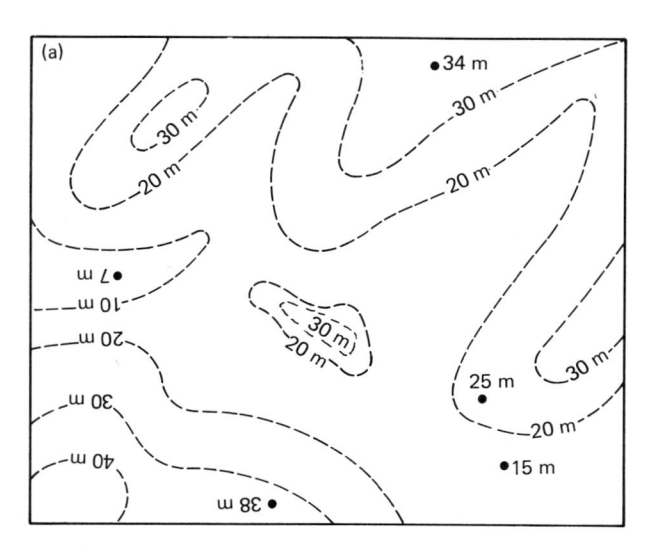

Fig. 10.1 Representations of relief

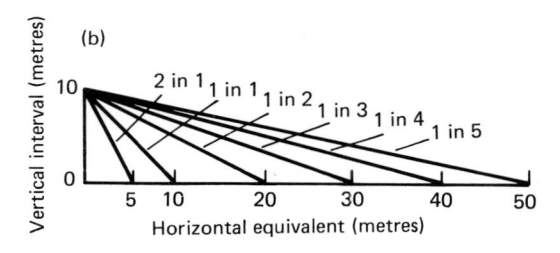

illustrated in Figure 10.1(*b*). The vertical axis of this graph represents a vertical contour interval of 10 m. The horizontal axis represents the *horizontal equivalent*. This is the distance between successive contour lines measured in a horizontal straight line (as on a map). In Figure 10.1(*b*) it is easy to see that the slope of the ground is greater when the contour lines are closely packed together and the horizontal equivalent is small. When the contour lines are 5 m apart (HE = 5) the average gradient is 2 in 1 (about 64°). When the contour lines are 50 m apart (HE = 50) the average slope is only a little over 10°. One way to gain an impression of the general relief features represented on a contour map is to draw a *profile* (section) or a number of profiles showing the rise and fall of the land surface. First of all the edge of a sheet of paper is laid across the map and the positions and heights of its intersections with the various contour lines are recorded. These points can then be joined up by a line which shows the form of the surface. Figure 10.2(*a*) shows a simple example of profiles which have been drawn for the equally spaced lines labelled A, B, C and D on the map. In Figure 10.2(*a*) all four profiles have been plotted on the same diagram. These are referred to as *superimposed profiles*. Each shows the rise and fall along one particular line. Taken together they give a summary of the general relief pattern depicted by the contours. When drawing such profiles it is a good idea to exaggerate the vertical scale so as to make the changes in gradient easier to see. In Figure 10.2(*b*) the profiles were first drawn in the way explained above, but then were regarded as being viewed from the south so that it was possible for the profiles to the north to be wholly or partially blocked from view by the ones to the south. Profiles arranged

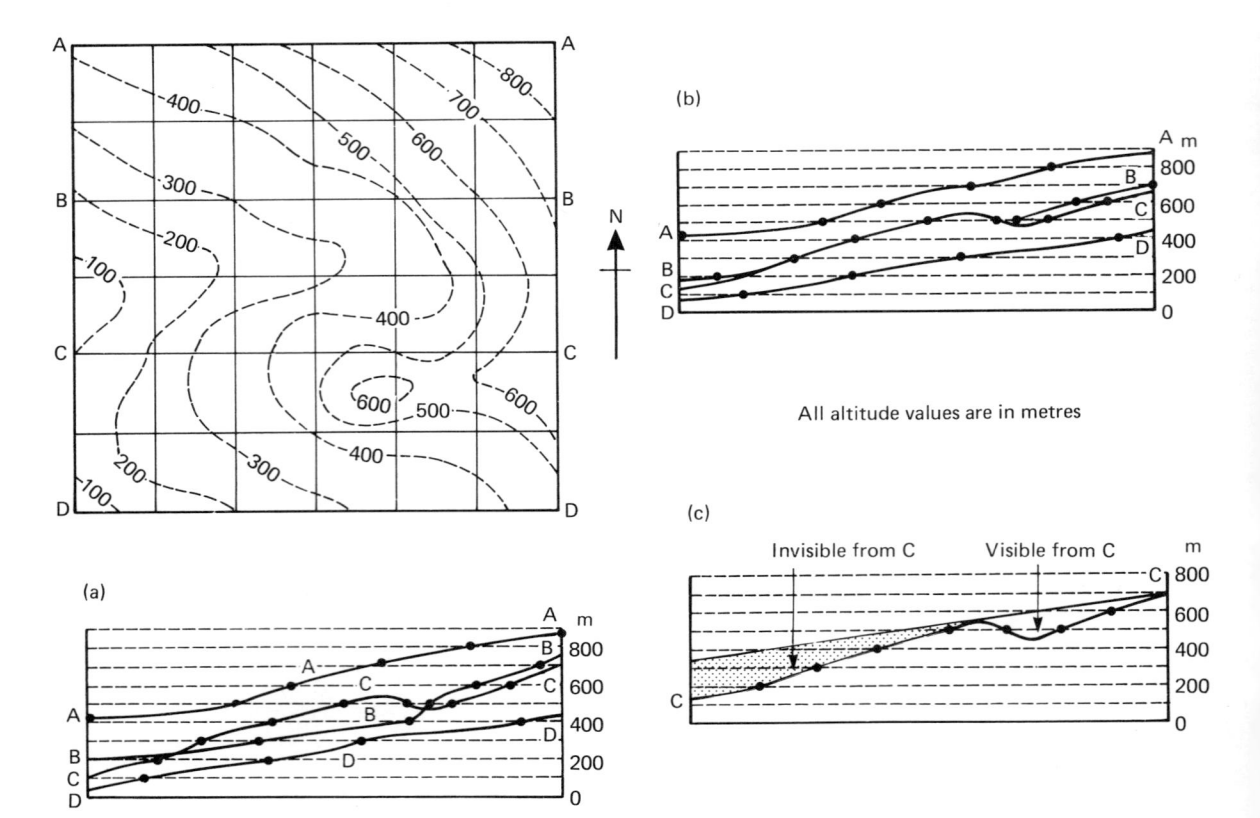

Fig. 10.2 Relief profiles

TOPOGRAPHICAL MAPS

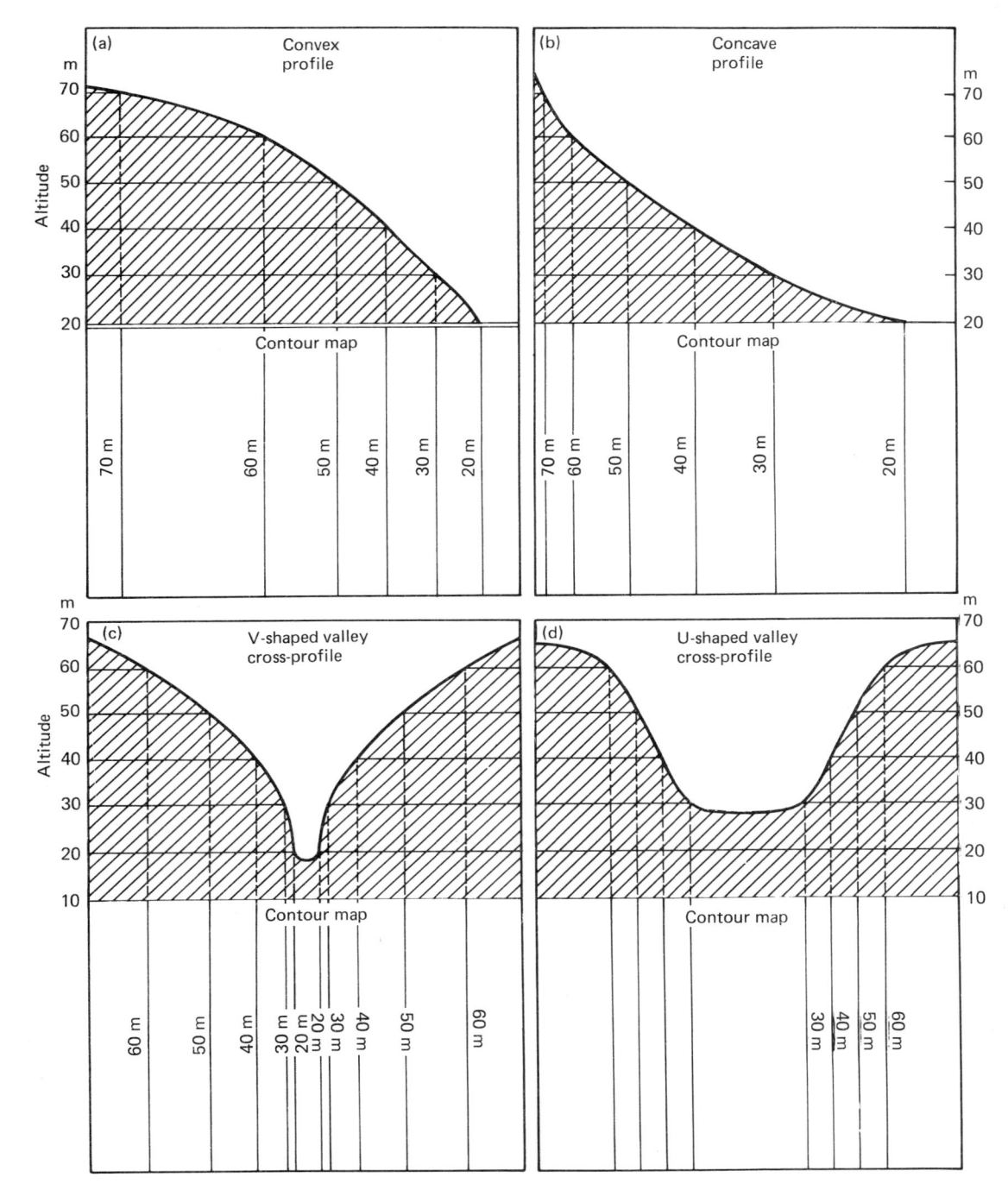

Fig. 10.3 Land form profiles (a)–(d)

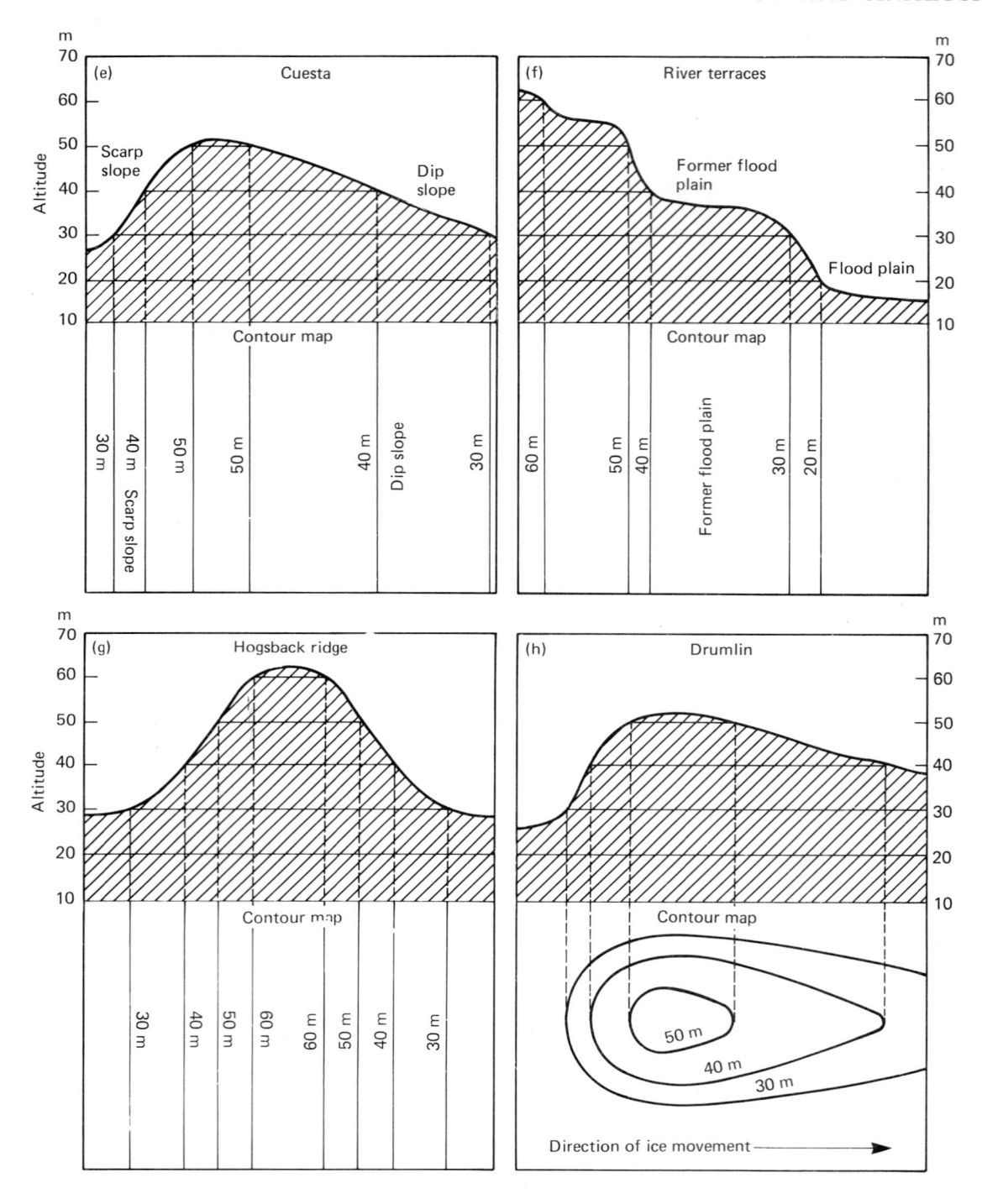

108 PRACTICAL GEOGRAPHY: PRESENTATION AND ANALYSIS

Land form profiles (e)–(h)

TOPOGRAPHICAL MAPS

in this way are called *projected profiles*. In this case however there is little difference between the superimposed and the projected profiles because most of the higher land is in the north. Profiles A, C and D were completely unaffected. Profile B however runs along an east–west trending valley for most of its length, so it is partly obscured by Profile C which runs along the side of the west-trending spur to the south.

Profiles can also be used to test *intervisibility* between different locations in an area. Profile C (Fig. 10.2(c)) provides a good example. Looking westwards from its eastern end the land at first falls a little. Then it rises to just over 500 m. This rise completely obscures all the rest of Profile C from view.

Landforms

Figure 10.3 shows how the profiles of various landforms are related to the contour patterns that represent them on maps. For the sake of easy comparison the diagrams are somewhat idealised. For example, most of the contour lines are shown as straight lines.

Figure 10.3(a) shows the characteristics of a convex hillside profile which becomes steeper downslope. Its associated contour map illustrates this in that the horizontal equivalent (Fig. 10.1(b)) between successive contour lines decreases in a downslope direction. In Figure 10.3(b) which has a concave hillside profile the contours become more widely spaced downslope in response to the reduced slope angles.

Figure 10.3(c) shows a valley with a V-shaped cross-profile such as may be found in the headwaters of an actively eroding river. The valley has steep sides, so the contours are fairly close together. Moving upslope on both sides of the river the slope angle slowly decreases and this is reflected in the slight but progressive increase in the spacing of contours. Figure 10.3(d) illustrates the general shape of the cross-profile of a U-shaped glaciated valley, with steep sides and a fairly flat floor. Few contour lines appear in the valley floor but they occur in quick succession on the valley sides.

Figure 10.3(e) shows the general characteristics of a cuesta. This landform is a fairly narrow ridge with a steep scarp slope on one side and a gentler dip slope on the other. A true dip slope has a slope angle equal to the angle of dip of the strata. It is easy to see that contours run generally parallel to the trend of the ridge and that contours are closer together on the scarp slope than on the dip slope.

River terraces (Fig. 10.3(f)) can be created when a river erodes alternately vertically and laterally. Lateral erosion can produce a wide and fairly level valley floor (a flood plain). Then the river may erode vertically and deepen its channel, and then laterally to create a flood plain at a lower level. The earlier flood plain thus becomes a river terrace. In this type of landform contours tend to be concentrated on the steeper slopes between the present flood plain and the lowest terrace and between river terraces at different heights.

A hogsback ridge (Fig. 10.3(g)) can be created when a weak rock layer is eroded on each side of a vertically dipping layer of resistant rock. The result is the creation of a usually fairly symmetrical, quite steep-sided ridge. Contour lines are spaced most closely on the slopes leading to the summit of the ridge.

Figure 10.3(h) shows the contour pattern typical of a drumlin, assuming that the ice which shaped it moved from left to right. A drumlin is usually steepest and highest at the end facing the direction from which the glacier approached.

10.2 Aspects of human geography

Topographical maps provide a great deal of information concerning human geography as well as physical geography. They show the characteristics of both urban and rural settlement patterns and to some extent the land-use distributions.

RURAL SETTLEMENTS

Elements of rural settlement patterns range from market towns to villages, to hamlets and isolated farms. Figure 10.4(a) shows an idealised distribution of villages in an area in which the relief features trend from east to west. In the south there is a flood plain. North of this the ground suddenly rises at the scarp slope of an east – west trending chalk cuesta (Fig. 10.3(*e*)) which rises to over 70 m. Further north the land surface begins to descend gradually northwards (dip slope). Three types of village site are located on the map. The *dry point site* is nearest the flood plain, but sufficiently high to escape most floods. The two *spring line villages* are located on the scarp slope at points where springs issue from the water table that underlies the cuesta. The two *dry valley villages* are sited on the floors of shallow dry valleys where a water supply can be obtained from wells which tap the underlying water table. This pattern of settlement is common in chalk areas such as the North and South Downs and the Chilterns. In this kind of area some villages, with advantages of accessibility will have grown into market towns, providing higher-order services for the villages. These are often located at nodal situations, such as a low-level gap across the cuesta or the bridging point of a large river.

Figure 10.4(*b*) shows a totally different kind of area. The contours indicate that there is a deep, steep-sided valley with a very narrow floor. The development of the settlement pattern has been related to the pattern of relief. The older housing and industrial buildings are crowded in the valley floor near the river. The housing is arranged in rows of small stone cottages. Industry was once dependent on water power from the stream. Since the 19th century the area has lost its comparative advantage in respect of manufacturing industry and the old mills have been adapted to accommodate modern industry, often of the 'service' type. The terraced houses still remain and have often been modernised. An estate of semi-detached houses has recently been built on the higher ground to the north-west where a main road gives access to the town's central business district. This kind of settlement pattern can frequently be seen in Lancashire, Yorkshire and South Wales.

URBAN SETTLEMENTS

Figure 10.4(*c*) shows the main features of a town and its road system. The original nucleus in the 19th century was situated at the crossing point of the two main roads. Here were located administrative offices, churches and a few resi-

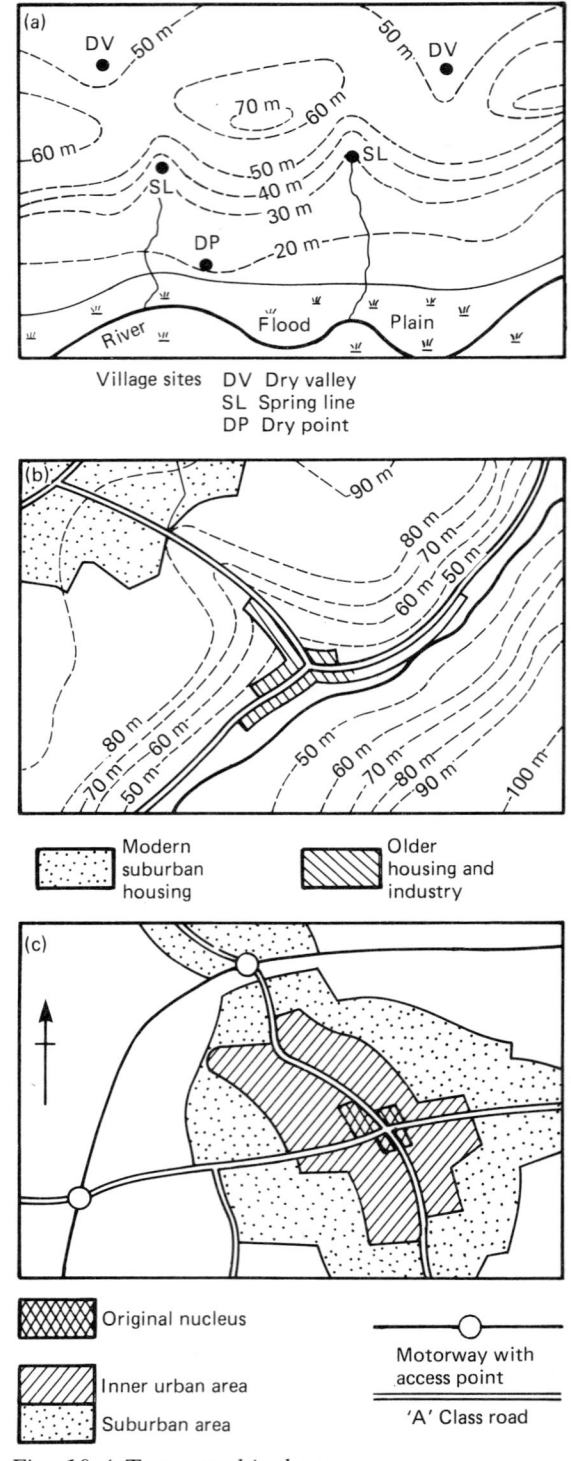

Fig. 10.4 Topographical maps

dences and workshops on the frontages of a pattern of narrow streets. Many of the older buildings have since been demolished and the original nucleus now functions as a shop and office zone (CBD). Through the years the town has expanded in a broad belt along each of its four main radiating roads and the inner area has come to be occupied mainly by manufacturing and service industries and terraced houses. Attempts are being made to 'renew' this inner area but much of it still remains rather shabby and some is quite derelict. The suburban area has developed more recently and is occupied by newer residences and industrial undertakings. Residents of this area create problems of traffic congestion as they travel to the town centre in the morning and home again in the evening. The latest development has been the building of the motorway through comparatively empty areas on the town's outskirts. This however does little to relieve congestion within the town, but it does allow relatively easy travel to and from neighbouring towns. It may not be possible to build a motorway from the suburban area to the CBD because of the amount of demolition that would be required.

Exercises

1. Refer to Figure 10.2.
   (a) For the equally spaced lines running from north to south across the map, and labelled P, Q, R and S, draw a set of:
   (i) Superimposed profiles;
   (ii) Projected profiles viewed from the east.
   (b) Comment on your diagrams.
   (c) Describe the position of the best location for viewing the area shown on the map. Explain your reasoning.
2. Refer to Figure 10.3. Draw contoured sketch maps to illustrate the following:
   (a) A river valley with a convex cross-profile in its upper reaches and a wide flood plain in its lower reaches;
   (b) A glaciated valley with a U-shaped cross profile and valley sides dissected by smaller valleys with V-shaped cross profiles;
   (c) A hogsback ridge changing laterally into a cuesta;
   (d) A conical hill whose slopes have been dissected by river erosion.
3. Using suitable Ordnance Survey maps draw generalised sketch maps of rural and urban areas similar to those in Figure 10.4.

11 Photographs

11.1 The interpretation of aerial photographs

Aerial photographs can be a great help in the study of the detailed geography of a relatively small area. Some aerial photographs have been taken with the camera pointing *vertically downwards*. Such photographs are very similar to large-scale maps. They usually cover only a small area compared with an Ordnance Survey sheet but they can show more geographical features than can be seen on a map. A map is necessarily confined to the representation of landscape features that can be surveyed accurately and which can be represented by lines or certain conventional symbols.

Figure 11.1(*a*) is a vertical aerial photograph of the approximate area shown in the map extract (Fig. 11.1(*b*)). It was taken some twelve years (1973) before the publication of the map (1985). Perhaps the most serious weakness of vertical aerial photographs is that they fail to portray relief features adequately. The photograph appears to be a level surface. The heights of physical features and buildings are not indicated at all. A map, on the other hand, depicts relief very efficiently by the use of contour lines. Also it is difficult to identify features on a black and white photograph. Roads, railways, trees and buildings are usually very clear but it is sometimes not so easy to identify lakes and reservoirs. The appearance of a landscape feature on an aerial photograph depends on the extent to which the feature reflects sunlight. Snow-covered areas and areas of limestone clints will tend to appear white. Water areas with no ripples on their surface will tend to look dark. Roads are represented by a light grey colour.

Figure 11.1(*a*) depicts an area at a scale that is slightly larger than the map extract (*b*), so it is quite easy to see the details of the area. The road pattern is quite clear. A main road enters the area at grid reference point 704160, just north of Dimple. At grid reference point 714142 this road branches at the War Memorial, one branch continuing south-eastwards and the other going towards the south. It is not difficult to identify the large industrial buildings near 718138 and the other group of industrial buildings to the south (717132), labelled 'Mills'. Also various areas of water can be identified in the far north-west of the photograph and in the Dunscar area.

An *oblique* aerial photograph can cover a much larger area than a vertical one. For an oblique photograph (Fig. 11.2) the camera is pointing at an angle to the ground rather than vertically downwards. On a vertical photograph the scale is almost constant over its whole area, so that it is similar to a map. In an oblique air photograph however the scale of the features depicted varies across the photograph. In the foreground at the lower margin of the photograph, its whole width is occupied by two buildings and a street. Near the photograph's upper margin there is room for a large industrial building on the left, and then, to the right, several blocks of flats, a bus station, a considerable length of road, a church and a large fruit and vegetable market. The boundaries of the area represented form a V-shape converging on the camera. Another difficulty in relating this photograph to a map of the area is that on the photograph much of the street plan is obscured by the height of the buildings.

11.2 The interpretation of ground-level photographs

Ground-level photographs can fill in details of the landscape which are not evident on the more generalised aerial photographs. To understand a ground-level photograph it is necessary to take into account the detailed features that are found on it. This may make it possible to deduce where the photograph was taken.

In terms of physical geography it is possible to formulate a broad classification of rock structures and their relationships with landforms. For example, ancient shields such as those of northern Canada and most of Africa are associated with plains or low plateaux with only an occasional high mountain. Ranges of fold mountains, such as the Alps, the Himalayas and the Andes have a more striking relief. Glaciated areas often have striking relief features even in Britain where the mountains are not very high. Limestone areas, sandstone areas, clay areas all have their specific landforms which help us to deduce the factors responsible for the relief characteristics shown on the photograph. It may also be possible to deduce the approximate location at which the photograph was taken.

Climate and vegetation (which may be dependent on climate) can also influence the character of an area shown on a photograph. A forested landscape is easily distinguished from grassland or desert. Detailed knowledge of common tree and plant types can help to locate the place where a photograph was taken.

Human characteristics of the population of the area may help to interpret a photograph. For example, house types vary somewhat over the world in response to varying characteristics of climate, available building materials and level of economic development. Various religions are associated with distinct types of church architecture, from the sometimes quite severe lines of the mosque (Islam) to the often highly decorated Buddhist temple.

Ground-level photographs can also provide information about the general level of economic development. In developing countries a simple, labour-intensive type of farming and industry contrasts with the mechanised type characteristic of more advanced countries. Animal-powered transport in developing countries contrasts with the emphasis on mechanised road transport in advanced countries.

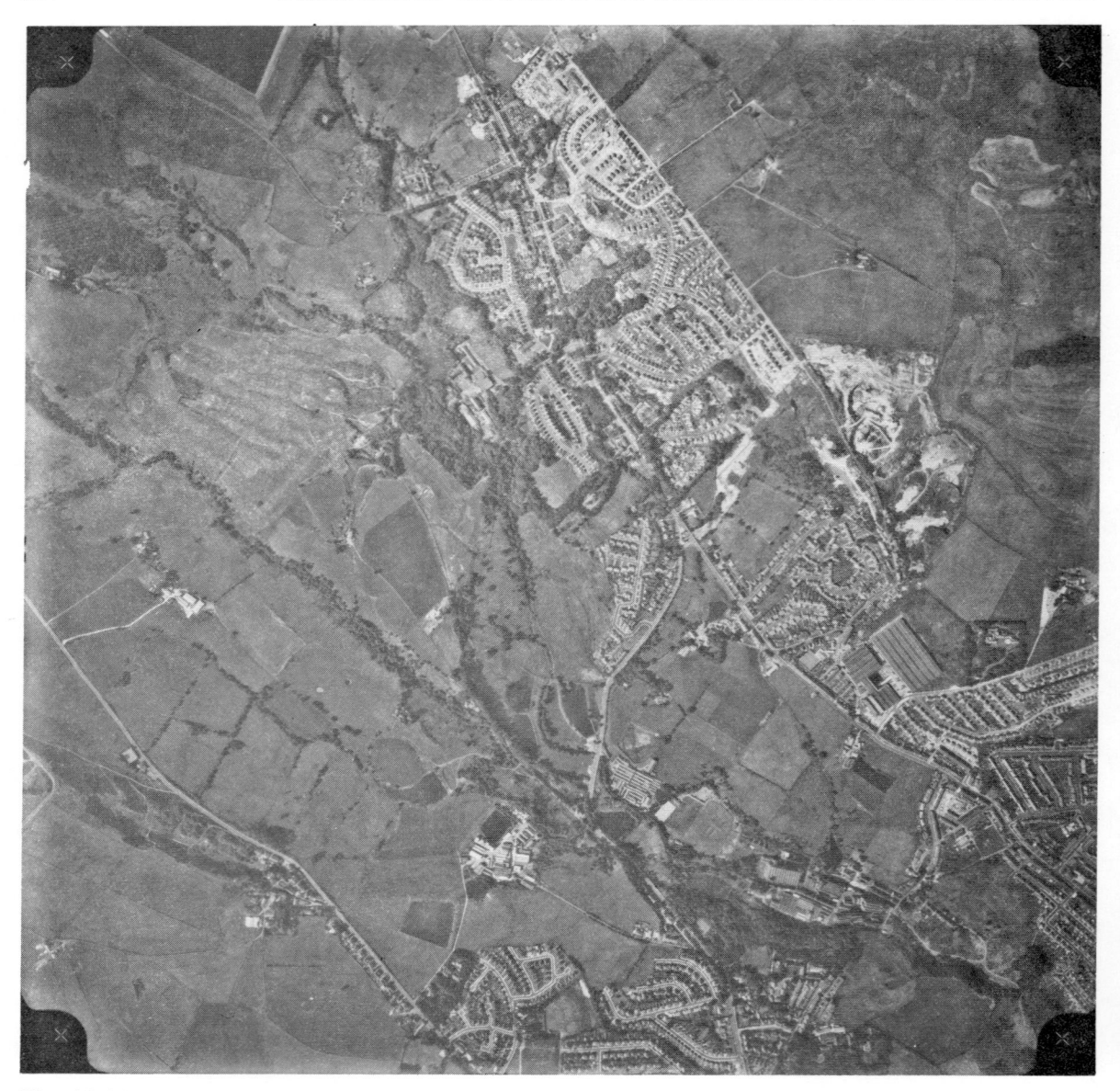

Fig. 11.1(a) Vertical aerial photograph

PHOTOGRAPHS

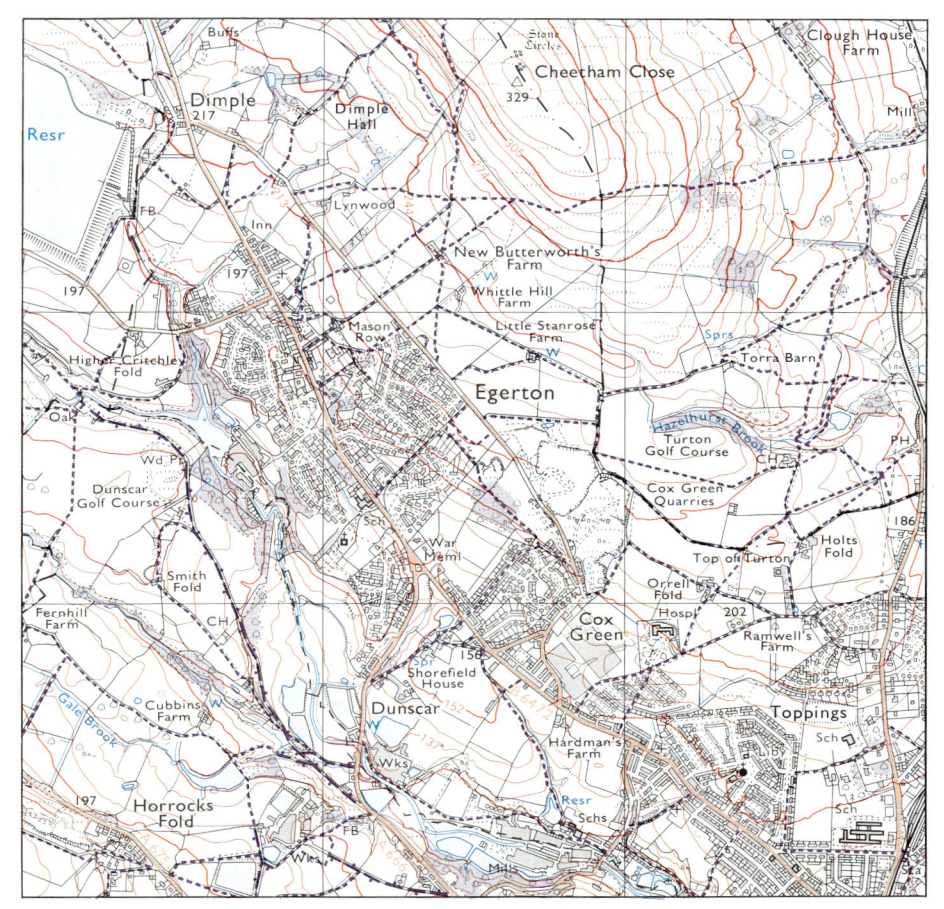

Fig. 11.1(b) Reproduced from the 1981 Ordnance Survey Pathfinder map with the permission of the Controller of Her Majesty's Stationery Office, Crown copyright reserved.

Fig. 11.2 Oblique aerial photograph

Exercises

1. The photographs show some of the characteristics of three urban areas (X, Y, Z) which are all situated between latitude 25°N and latitude 45°N.

   (a) Using the photographs discuss the contrasting geographical characteristics of these three urban areas.
   (b) Suggest a location for each urban area and explain the reasons for your choice.

Urban X

Urban Y

Urban X

Urban Y

Urban Z

PHOTOGRAPHS

11A

11C

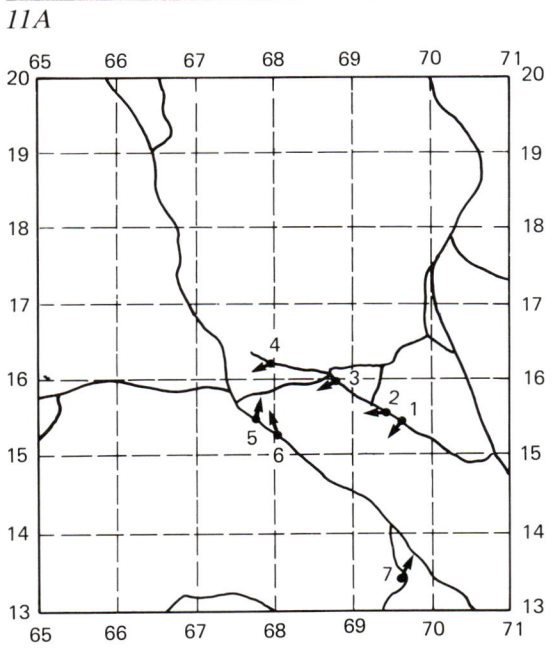

Viewpoints

11D

11B

11E

11F *11G*

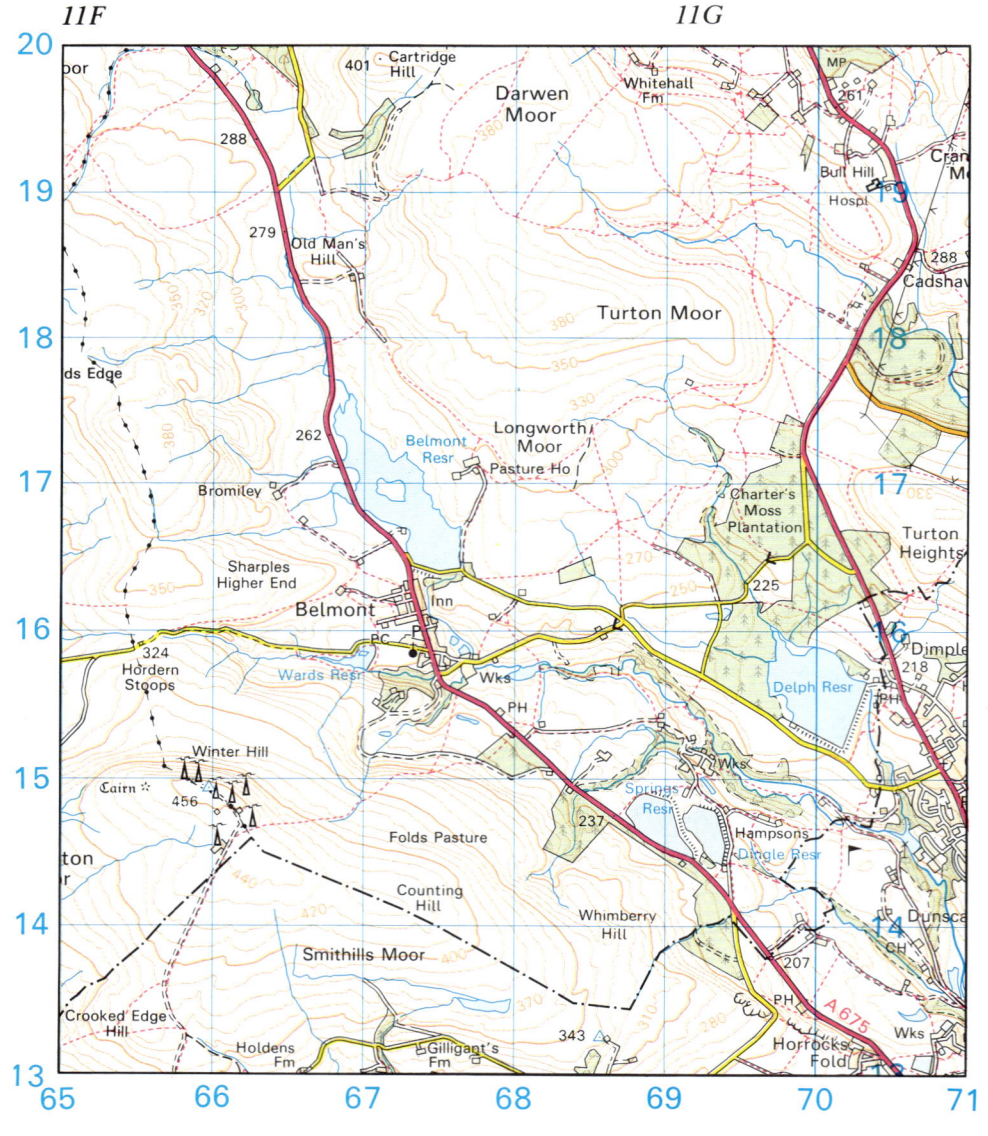

PHOTOGRAPHS

2. Refer to Figure 11.1(*a*) and (*b*).
   (a) Describe the changes in the settlement pattern that have taken place in the period between the date that the photograph was taken and the date of the map survey.
   (b) To what extent does the development of the area shown on the map extract appear to have been influenced by physical geography?
3. Look at pages 117 and 118. The sketch map shows the road network of the area represented by the Ordnance Survey map extract. It indicates the location of the points selected for taking photographs to illustrate the characteristics of the area. It also shows the direction in which the camera was pointing at each of these locations. The viewpoints are numbered from 1 to 7 and the pictures are labelled 11A to 11G.
   (a) Study the photographs and the maps and then state which of the pictures (11A to 11G) was taken at each of the viewpoints (1 to 7).
   (b) Study all the material available and then write a concise account of the geographical characteristics of this area.

Reproduced from the 1983 Ordnance Survey 1:50 000 Landranger map with the permission of Her Majesty's Stationery Office, Crown copyright reserved.

Index

Accessibility, 87–90
Accumulated temperatures, 15–17
Air masses, 99–100
Alternative hypothesis, 96
Anticyclone, 102–03
Array, 1
Associated number, 87
Averages, 61–4

Bar chart, 6–8
Basin relief, 81
Best fit line, 31–2
Bifurcation ratio, 83–4

Chi-square analysis, 74–7
Choropleth maps, 37–41
Circular graph, 9–11
Classification of climates by means of a matrix, 58
Coefficient of variation, 67
Combining two line graphs, 14
Contiguity of countries in Western Europe, 59
Correlation coefficient, 95

Depression, 100–2
Desire line maps, 53–4
Dispersion index, 87
Dot maps, 36–7, 40
Drainage density, 81
Drainage networks, 81–4

Flow diagrams, 50–3
Flow line maps, 50–3
Form factor, 80
Frequency polygon, 4–5
Fronts, 100–2

Graph theory, 84–6
Graphs with logarithmic scales, 21–6
Ground level photographs, 112-3

Histogram, 2–4
Horizontal equivalent, 105
Hypsometric curve, 17–8

Isoline maps, 44–50
Isotims and isodapanes, 46-9

Landforms, 109
Length ratio, 84–5
Line graph, 11–31
Location quotient, 92–4
Lorenz curve, 19–21

Mean, 61–2
Mean centre, 71
Mean deviation, 65–6
Measures of dispersion, 65
Median, 62–4
Median centre, 71
Mode, 64
Moving average, 13

Nearest neighbour analysis, 77–9
Normal distribution, 68–70
Null hypothesis, 96

Oblique aerial photograph, 113–5
Ogive, 5–6

Pie graph, 9
Principal diagonal, 57
Proportional symbols, 41–3

Quartile, quartile deviation, 65

Range, 65
Rank correlation, 95
Regression line, 31–2
Relief, 105
Running mean, 13–14
Rural settlements, 109–10

Scatter diagrams, 26–9
Semi-interquartile range, 65
Shape index, 81
Shimbel index, 87

Significance level, 96
Skewness in a distribution, 68
Standard deviation, 66–7
Standard distance, 73
Stream channels, 81–2
Stream orders, 82
Superimposed profiles, 106
Symmetric matrix, 57

Time accessibility, 89–90
Topology, 84–6
Transport networks, 84–9
Triangular graph, 26–9

Urban settlements, 110–1

Vertical aerial photograph, 114–5